Palaeopathology in Egypt and Nubia

A century in review

Edited by

Ryan Metcalfe, Jenefer Cockitt
and Rosalie David

Archaeopress Egyptology 6

Archaeopress
Gordon House
276 Banbury Road
Oxford OX2 7ED

www.archaeopress.com

ISBN 978 1 78491 026 6
ISBN 978 1 78491 027 3 (e-Pdf)

© Archaeopress and the individual authors 2014

All rights reserved. No part of this book may be reproduced, stored in retrieval system, or transmitted, in any form or by any means, electronic, mechanical, photocopying or otherwise, without the prior written permission of the copyright owners.

This book is available direct from Archaeopress or from our website www.archaeopress.com

Printed and bound in Great Britain by Marston Book Services Ltd, Oxfordshire

In loving memory of Judy Miller, member of The Manchester Egyptian Mummy Team, and esteemed colleague and friend

Preface

The original inspiration for this volume came from research that focussed largely on the work of Sir Grafton Elliot Smith and the Archaeological Survey of Nubia. These combined areas of study present an intimidating volume of work, but they cannot be fully understood without also looking at the evolution of the disciplines of palaeopathology and bioarchaeology. Elliot Smith and other early pioneers played pivotal roles in the foundation of these subjects. In many instances they were able to produce studies that have remained unmatched in terms of their scale and impact, and the collections of human remains they amassed continue to present valuable research opportunities.

Both the conference that inspired this volume, and the chapters contained herein provide an opportunity to see how these disciplines have changed over the last hundred years, and explore the directions we may expect them to take in the future. New technologies and improvements to existing methods are being constantly developed and applied to archaeology, providing scholars and scientists with truly exciting opportunities for innovative projects.

The editors would like to dedicate this volume to Dr Judy Miller, teacher, colleague and friend to so many of the KNH team over many years, and who very sadly passed away during its preparation.

Ryan Metcalfe, Jenefer Cockitt and Rosalie David

Manchester, August 2014

Acknowledgements

The editors would like to thank the following for their support:

The Wellcome Trust who sponsored the workshop 'Palaeopathology in Egypt and Nubia: A century in review' where these papers were first presented and discussed, and who supported the research project 'Sir Grafton Elliott Smith and the Archaeological Survey of Nubia: their significance to the palaeopathological tradition,' which is the focus of much of this volume;

The Natural History Museum, London for hosting the workshop and for their support and enthusiasm throughout the duration of the project and the production of this book;

The K.N. Hinckley Charitable Trust which has continued to support the work of researchers and students at The University of Manchester;

All the authors and reviewers for their invaluable contribution to this volume and for their patience and support throughout the editorial process.

Contents

Preface ii

Acknowledgements iii

Chapter 1: History of bioarchaeology

Sir Grafton Elliot Smith: Palaeopathology and the Archaeological Survey of Nubia 1
Rosalie David

Whose body? The human remains from the 1908-1909 season of the Archaeological Survey of Nubia 9
Jenefer A. Cockitt

The more things change? The archaeological work of Alfred Lucas 23
Ryan Metcalfe

Chapter 2: Palaeopathology

Harris lines, ill health during childhood, poor diet, emotional stress or normal growth patterns? 31
Abeer Eladany

An interesting example of a condylar fracture from ancient Nubia suggesting the possibility of early surgical intervention 41
Mervyn Harris, Tristan Lowe and Farah Ahmed

An overview of the evidence for tuberculosis from ancient Egypt 51
Lisa Sabbahy

Palaeopathology, disability and bodily impairments 57
Sonia Zakrzewski

Chapter 3: Dental palaeopathology

Dental infections in ancient Nubia 69
Roger J. Forshaw

A case of severe ankylosis of temporomandibular joint from New Kingdom necropolis (Saqqara, Egypt) 83
Ladislava Horáčková and Frank Rühli

Occlusal macrowear, antemortem tooth loss, and temporomandibular joint arthritis at Predynastic Naqada.. 95
Nancy C. Lovell

Chapter 4: Mummification

How to make a mummy: A late hieratic guide from Abusir 107
Jiří Janák and Renata Landgráfová

Studying mummies: Giving life to a dry subject 119
Michael R. Zimmerman

Chapter 5: Imaging in bioarchaeology

Microstructural analysis of a Predynastic iron meteorite bead...............................129
Diane Johnson, Monica M. Grady, Tristan Lowe and Joyce Tyldesley

Imaging and analysis of skeletal morphology: New tools and techniques................141
Norman MacLeod

Chapter 6: Digital resources

Mummies on rails..157
Ahmad Alam, Ian Dunlop, Robert Stevens, Andrew Brass, Jenefer Cockitt, Rosalie David and Ryan Metcalfe

Mummy website and database..167
Barbara Zimmerman, Sukeerthi Shaga, Pavitra Kaveri Ramnath, and Sai Phaneendra Vadapalli

List of Figures and Tables

Whose body? The human remains from the 1908-1909 season of the Archaeological Survey of Nubia
    Table 1: The provenanced bones from the ASN 12
    Table 2: Skulls from the Archaeological Survey of Nubia 13
    Table 3: The surviving human remains from the Nubian Pathological Collection 13
    Table 4: The provenanced bones from the ASN 15
    Figure 1: An example of a surviving ASN tomb card from cemetery 87, grave 84. 16
    Figure 2: Top – The top of two skulls from 74/514 and 76/83 ;
    Bottom – The faces of two skulls from 87/90 and 87/131 18
    Figure 3: Photograph B763 excavated skeleton in grave 19

An interesting example of a condylar fracture from ancient Nubia suggesting the possibility of early surgical intervention
    Figure 1: Condylar head showing extensive destruction of the normal anatomy 43
    Figure 2: View looking directly into the Glenoid fossa 44
    Figure 3 Hypertrophic bone of mandibular. 44
    Figure 4: Demonstrating a sharp incision 45
    Figure 5 A and B: Alicona surface laser scan 46
    Figure 6: 3D reformation of Micro CT 46
    Figure 7: Anterior view of skull with mandible articulated. 47
    Figure 8: Occlusal surface of molar teeth demonstrating even occlusal wear 48

Dental infections in ancient Nubia
    Figure 1: Skull NU363 displays a periapical cyst 73
    Fig. 2 Occlusal view of skull NU363 74
    Figure 3: Multiple pathological bony cavities are evident in skull NU131. 74
    Figure 4: Skull NU322 displays a cyst associated with the second maxillary premolar. 75
    Figure 5: A cyst measuring 9 x 20 mm 76
    Figure 6: Periapical lesion above the roots of the first molar in skull NU616. 76
    Figure 7: Diagram representing the maxillary air sinus. 77
    Figure 8: Skull NU737 is from one of the individuals excavated 78
    Figure 9: Palatal view of the cyst in skull NU737 79

A case of severe ankylosis of temporomandibular joint from New Kingdom necropolis (Saqqara, Egypt)
    Figure 1: Fragments of the skull of a 16-17 years old individual 87
    Figure 2: Mandible of a 16-17 years old individual from the north chapel of tomb 87
    Figure 3: Porous and rough bottom of the left mandibular fossa 88
    Figure 4: View to the left temporomandibular joint with ankylosis. 88
    Figure 5: The affected right temporomandibular joint. 89
    Figure 6: X-ray photograph of affected temporomandibular joint 91

Occlusal macrowear, antemortem tooth loss, and temporomandibular joint arthritis at Predynastic Naqada
    Table 1: The Naqada skeletal sample of adults available for examination of the teeth 96

    Table 2: Patterns of occlusal macrowear at Naqada (tooth count). 99
    Table 3: Tooth count and individual count patterns of antemortem tooth loss 99
    Table 4: The age, sex, and cemetery distributions of TMJ lesions at Naqada 100
    Figure 1: Resorption on the right articular eminence 101
    Figure 2: Extensive resorption and marginal lipping 102
    Figure 3: Pronounced remodelling of the right glenoid fossa 103
    Figure 4: Proliferation of bone within the left glenoid fossa.. 104

How to make a mummy: A late hieratic guide from Abusir

    Table 1: Summary of late hieratic inscriptions from the embalmer's cache 108
    Table 2: Substances sorted by days of the mummification process. 114
    Table 3: Comparison of the evidence of texts 115

Microstructural analysis of a Predynastic iron meteorite bead

    Figure 1: Tomb card of tomb 67 Gerzeh cemetrery 131
    Figure 2: Iron beads found in Gerzeh cemetery tomb 67, ... 132
    Figure 3: The Gerzeh bead analysed in this study shown as it is today, ... 133
    Figure 4: Secondary electron SEM image of hydrated iron oxide growth structures 134
    Figure 5: Secondary electron SEM image of one end of the bead showing exposed fibres . 135
    Figure 6: Image of Gerzeh bead CT model oxide layers 136
    Figure 7: Gibeon, a coarse octahedrite iron meteorite, ... 137
    Figure 8: Optical image of the prehistoric Egyptian Gerzeh bead 138

Imaging and analysis of skeletal morphology: New tools and techniques

    Figure 1: Standard human cranial landmarks and measurements. 142
    Figure 2: A. Digital callipers, used for measuring 3D distances between landmark points .. 144
    Figure 3: A. d'Arcy Thompson's drawing of a transformation grid 145
    Figure 4: Three different types of modern 3D digitisers. 147
    Figure 5: Steps in sampling surface morphology using the eigensurface procedure 148
    Table 1: A small sample of mixed male and female crania 149
    Figure 6: Theory behind use of pixel coordinate brightness/colour values 150
    Figure 7: Principal component ordination of the specimen images shown in Table 1. 151

Mummies on rails

    Figure 1: The MEPR main record list screen.. 163

Mummy website and database

    Figure 1: The main screen used for searching the database... 168
    Figure 2: An example search result. 169
    Figure 3: An example of the range of data available for a single slide.................................. 169

Sir Grafton Elliot Smith: Palaeopathology and the Archaeological Survey of Nubia

Rosalie David

KNH Centre for Biomedical Egyptology, The University of Manchester, Manchester, UK

Abstract

A three-year study, 'Sir Grafton Elliot Smith and the Archaeological Survey of Nubia: their significance to the palaeopathological tradition,' was undertaken (2010-2013) at The University of Manchester in partnership with The Natural History Museum, London. This has attempted to assess and redefine Sir Grafton Elliot Smith's contribution to palaeopathology, focusing on his role as Anthropological Advisor to the Archaeological Survey of Nubia. It has also addressed the current state of preservation of the skeletal/archaeological collection from this major rescue project. The preliminary results of this research were presented at a Workshop, 'Palaeopathology in Egypt and Nubia: A Century in Review' (August 29-30, 2012), held at The Natural History Museum, London. This paper provides a brief resumé of Elliot Smith's life and career as an anatomist and palaeopathologist, setting his work in the context of early studies and the development of palaeopathology. It also outlines the main aims and objectives of the recent study.

Early 'unrollings' of mummies

Ancient Egypt became a focus of interest for antiquarians and collectors from Renaissance times onwards, and museums, learned societies, and wealthy individuals in Britain, Europe, and the United States of America vied to purchase antiquities that would enhance their growing collections (Wortham, 1971). Mummies, sometimes regarded as unusual souvenirs of a tour to Egypt, were often amongst these early acquisitions, and from the sixteenth century, many were 'unrolled' (unwrapped) to provide a highlight for an audience invited to attend a social evening. These were frivolous events, and the unwrappings had little scientific value; in most cases, no record was kept or has survived, although occasionally, renowned investigators led the procedure and their published results preserve reliable evidence.

Nevertheless, some of the unwrappings undertaken in the eighteenth and nineteenth centuries were examples of good scientific practice, the projects being undertaken either by multidisciplinary teams, or by serious individual researchers. Leading authorities included Thomas J. Pettigrew (1791-1865), a London surgeon and antiquary who undertook many 'unrollings' and published a history of Egyptian mummies (Pettigrew, 1834); and Augustus B. Granville (1783-1872), a British physician of Italian origin with a great interest in Egyptian mummies who published an important account of ovarian disease in an Egyptian mummy (Granville, 1825). Significant interdisciplinary projects included those undertaken on mummies owned by learned societies: for example, the Belfast Mummy in 1835, and the Leeds Mummy in 1825 (Osburn, 1828).

Beginnings of palaeopathology and the scientific method

Unscientific unrollings were discontinued in the middle of the nineteenth century, and by the early twentieth century, various pioneers were laying the foundations for 'mummy science'. At The University of Manchester (UK), Dr Margaret Murray's interdisciplinary team unwrapped, autopsied, and subsequently investigated and produced a scientific report on the mummies and tomb goods belonging to the 'Two Brothers' (Murray, 1910). Armand Ruffer (1859-1917), a British pathologist who became Professor of Bacteriology at Cairo Medical School in Egypt, devised methods of rehydrating ancient tissues, and 'Ruffer's solution' is still used to rehydrate mummified tissue prior to microscopic examination. He also developed histological techniques to identify disease in Egyptian mummies and skeletons, and key aspects of his work were published posthumously (Ruffer, 1921). He pioneered the study of disease in ancient populations as a distinct and separate area of science, and this sub-discipline, for which he invented the term 'palaeopathology', remains highly relevant to the history of science and medicine, facilitating the study of antiquity through ancient human remains, and providing a basis and context for modern disease studies.

Alfred Lucas, O.B.E. (1867-1945), was a British chemist who also carried out pioneering research in this field. First going to Egypt in the hope of alleviating a lung condition, he became Chemist to the Government Salt Department in Cairo, and Chemist to the Antiquities Service (1923-1932). He played a leading role in the analyses of ancient materials and substances, publishing a uniquely important account of his experiments and observations (Lucas, 1962). He also pioneered conservation and restoration techniques for cleaning and treating excavated material, notably supervising nine years' work undertaken on the contents of the tomb of Tutankhamun.

The legacy of Grafton Elliot Smith

The contribution made to Egyptology and palaeopathology by another early scientist, Sir Grafton Elliot Smith (1871-1937), is the subject of the current project entitled "Sir Grafton Elliot Smith and the Archaeological Survey of Nubia: their significance to the palaeopathological tradition" (supported by The Wellcome Trust [WT090575MA]). This project had two main aims: by examining the importance of Elliot Smith's role as anthropological advisor to the Archaeological Survey of Nubia (ASN), to reassess and acknowledge his hitherto unrecognised contribution to the development of palaeopathology, and to highlight the research potential of the unrivalled skeletal collection obtained from the ASN.

Until now, Elliot Smith's study of the Egyptian royal mummies has been his most acclaimed contribution to palaeopathology. While he was Professor of Anatomy in the Cairo School of Medicine (1900-1909), Gaston Maspero, the Director-General of Antiquities in Egypt, invited him to examine two caches of royal mummies that had been discovered at Deir el-Bahri (1881) and in the Theban tomb of Amenhotep II (1898). This extensive study of the mummies of the rulers of the New Kingdom (1567 BCE – 1085 BCE) ultimately formed the basis of his classic account (Smith, 1912). Over a period of two years, with additional access to the mummies of priests and commoners, Elliot Smith was able to identify various embalming methods and techniques of mummification in use at different periods, which provided him with the material for a pioneering publication (with co-author, Warren R. Dawson) on techniques of mummification (Smith and Dawson, 1924).

However, Elliot Smith's most groundbreaking contribution is arguably the research which he undertook, with his co-workers Douglas Derry and Frederick Wood Jones, on the thousands of Egyptian and Nubian mummies and skeletons recovered during the Archaeological Survey of Nubia. This was a heritage rescue operation established to save some of the ancient remains threatened by construction of the first dam on the Nile at Aswan in the early 20th century (Smith and Jones, 1910).

Grafton Elliot Smith: The early years

Grafton Elliot Smith was an Australian anatomist and anthropologist with diverse interests and achievements (Dawson, 1938; Crook, 2012). In the medical sphere, his significant contributions include groundbreaking research on the human brain, and innovative methods that changed the course of anatomy teaching in medical schools. In his day, he was a foremost authority on the evolution of man, and a leading proponent of the 'Diffusionist Theory' (the claim that all the world's cultures had emanated from a small number of great civilisations, most notably Egypt), a belief for which he was later criticised and ridiculed.

Named after the town – Grafton in New South Wales, Australia – where he was born, Elliot Smith was the son of a London-born country school-master and his wife, Mary Jane Evans. He recorded that it was his father who initiated his scientific interest, encouraging him to collect specimens of flora and fauna (Smith, 1938, p.114):

> "...when on vacation visits to the seaside I found the carcase of a dead shark on the beach, I proceeded to dissect it with a penknife, and became specially intrigued by the brain which seemed to me to be a veritable collection of puzzling tricks."

From the age of ten, the boy's enthusiasm for physiology was well established. This led him to a career in medicine, and in 1892, he graduated from Sydney University with a Bachelor of Medicine degree. In 1896, with the award of a university travelling scholarship, he set out for London. His studies brought recognition when he was awarded a Research Studentship at St John's College, Cambridge, where his academic progress was further influenced by the Professor of Anatomy, Alexander Macalister. Ernest Rutherford (who, in 1908, was awarded the Nobel Prize for Chemistry) was Elliot Smith's contemporary at Cambridge, and later described him in the following terms (Rutherford, 1938, p.134-135):

> "At the time of which I speak, Elliot Smith was just twenty-five years of age. He was shy and taciturn at first with strangers, while his drooping moustache gave him an appearance almost of melancholy. This soon vanished when he talked with friends on matters in which he was interested, when he became lively and humorous and the best of company. His outward appearance changed markedly in middle age, when he

was clean shaven, rubicund, and a ready and fluent speaker, looking, to my mind, rather like a distinguished and jovial bishop."

Elliot Smith's introduction to Egyptology

Macalister next arranged for Elliot Smith to become the first Professor of Anatomy at the new Government Medical School in Cairo, a post he held from 1900 until 1909. During this period, his anatomical work brought him into contact for the first time with the ancient human remains that archaeologists were uncovering. His initial involvement with Egyptology came about in 1902 when, because of his special interest in the human brain, he was asked to examine a series of ancient human brains from the site of El Amrah (Smith, 1902).

Elliot Smith's developing expertise in this area soon led to an invitation from Gaston Maspero, the Director-General of Antiquities, to participate, with the archaeologist Howard Carter, in the investigation of a mummy in the presence of the pro-Consul, Lord Cromer. Maspero then asked Elliot Smith to examine the mummy of Tuthmosis IV which had been discovered in 1898. In order to provide a detailed anatomical evaluation, Elliot Smith decided to use a new analytical facility – radiography. The only machine available in Cairo was located in a private nursing home: the mummy, accompanied by Elliot Smith and Howard Carter, was transported in a horse-drawn cab, and the first radiographic study of a full-body human mummy was undertaken (Smith, 1912, p.iii-iv, vi-vii).

The Archaeological Survey of Nubia

A decision to raise the first dam at Aswan, with the prospect of resultant flooding of antiquities in that area, prompted Maspero, in 1907, to establish the first Archaeological Survey of Nubia. Placed under the directorship of the American Egyptologist, George Reisner, this was the first extensive archaeological rescue project in the area, and ultimately completed the excavation of over 20,000 burial sites. The results were published in a series of bulletins and reports (e.g. Reisner, 1910; Smith and Jones, 1910), and its contribution to palaeopathology remains unrivalled. As well as acting as a rescue mission, the ASN also sought information on the pattern of successive races and racial mixtures in the area; the extent of the population at different periods; the source and degree of civilisation; the economic basis of existence; and the character of industrialised products. The ASN examined 151 cemeteries in five years; in the first two months alone, researchers worked on eleven cemeteries containing more than 3,000 bodies. This work, covering a period from Predynastic (A-group) to Medieval times, revealed graves, tombs, temples, grave goods, skeletal remains, mummies and animals.

Elliot Smith accepted an invitation to act as anthropological advisor, with responsibility for the ASN's anatomical reports, but he could not have envisaged the scale of the work involved in exhuming and examining thousands of human, as well as animal, remains. Assistants were appointed – from 1907, Douglas Derry, and Frederick Wood Jones. Together with Elliot Smith, they undertook a systematic study of osteological material and statistical analysis, producing extensive data about disease and trauma in this population. They pioneered modern epidemiological research with this first large-scale study of disease patterns of particular populations (in this case, ancient Egypt and the Sudan). Wood Jones provides an intriguing assessment of Elliot Smith's attitude to work (Jones, 1938, p.139):

> "...more than any other man I have ever met, he was indifferent to his physical surroundings....it might be said with truth that he carried his own environment with him....The only local incidents that affected him were the material objects, such as anatomical subjects, libraries and museums, that happened to be within his reach.
>
> It is certain that his Egyptian period changed and enlarged his outlook; but it was not, as many have supposed, because the romance of the land of the Pharaohs attracted him or had him under a spell. It was rather because Egypt furnished him with skulls and skeletons and mummies; and upon these things he was asked to report to Egyptologists."

Elliot Smith in Manchester and London

Elliot Smith left Cairo in 1909 to take up a post as Professor of Anatomy at The University of Manchester. He remained there until 1919, playing a leading role in creating a first-class School of Anatomy. At this time, Manchester was at the centre of intellectual and scientific interaction, and had attracted leading specialists in a number of fields; these included Rutherford, with whom Elliot Smith was able to renew his friendship.

A collection of human skeletal material, originally owned by Elliot Smith, is still held at The University of Manchester. These remains were examined by a scientist some years ago who classified them as a teaching collection that had no association with the ASN. However, two sharp-eyed postgraduate students engaged in cataloguing the collection noticed excavator's notation marks on some of the bones, and this discovery later enabled some of the skulls and femora to be attributed to the ASN. A pilot scheme which then placed this material into its true context, and enabled it to be tied into the remainder of the known surviving ASN collections, provided the basis for the current project.

Elliot Smith moved to London and completed his career at University College, where he held the Chair of Anatomy from 1919 to 1936. He was involved in the College's strategic plan to integrate biological and social sciences with humanities, supported by funding of over £1.2m, donated by the Rockefeller Foundation in 1920. However, because of funding issues, exacerbated by the 1929 world depression, this ambitious plan did not materialise, and although Grafton Elliot Smith developed cutting-edge anatomy and research-based anthropology, his own vision for an Institute of Human Biology as part of the overall scheme was never fully realised.

The current project: 'Sir Grafton Elliot Smith and the Archaeological Survey of Nubia: their significance to the palaeopathological tradition'

The ASN was one of the earliest, most extensive archaeological rescue projects ever undertaken: human and animal remains and associated archaeological material were retrieved from hundreds of sites in southern Egypt, threatened by rising water levels. Subsequently, the reclaimed skeletal remains, artefacts and records were distributed to institutions around the world. But why is there now a need to revisit the ASN? First, although the material is of unique significance to disease and anthropological studies, its potential has never been explored, and the original conclusions remain unpublished. The distribution of the ASN material to centres around the world has resulted in fragmented collections, where the physical remains and artefacts are sometimes separated from the archives; cultural interpretation is lacking; and there has been an insignificant amount of modern epidemiological research on this unparalleled material. There is an urgent need to identify and record this unique evidence.

This three-year study (2010-2013) study was undertaken by the KNH Centre for Biomedical Egyptology at The University of Manchester in partnership with The Natural History Museum, London, which holds a major part of the ASN collection. Collaborators on the project included The Duckworth Collection, University of Cambridge; The Manchester Museum; the National Research Centre, Cairo; and in the United States of America, the renowned palaeopathologist, Professor Michael Zimmerman. Other research associations were established with The Royal College of Surgeons, London; The Royal Society of Medicine; the British Museum; the Boston Museum of Fine Arts (USA); and institutions in Egypt.

Archival and scientific resources were used to assess Elliot Smith's contribution to palaeopathology, and the current state of preservation of the skeletal/archaeological collection. Also, radiographs and other techniques were employed to take forward the research on disease patterns that was initiated by Elliot Smith and his colleagues. Current locations of the ASN collections were traced and the material reunited on a publicly accessible, dedicated website as a research resource which can now provide information about identification and distribution locations of excavated materials, disease occurrence, trauma, diet and so forth. In due course, the website will include verification of Elliot Smith's disease diagnoses, results of new disease studies, and statistical analyses of disease patterns based on the collated data.

The results of the project have been made available in academic and popular publications. Outreach activities based on the study have included a day-school in Manchester; an exhibition, "Grave Secrets", at The Manchester Museum (November 2011 – March 2012); and, at The Natural History Museum, London, a Workshop entitled 'Palaeopathology in Egypt and Nubia: A Century in Review' (August 29-30, 2012), and a Public Lecture (August 28, 2012) by Professor M. R. Zimmerman: 'Studying Mummies: Giving Life to a Dry Subject'.

Elliot Smith died on January 1st, 1937. His ideas were wide-ranging and sometimes controversial, but he always believed in a rigorous, scientific review of facts and their interpretation (Crook, 2012, p.151, note 19):

> "We must particularly guard against ideas becoming so embedded that they are accepted as gospel without check or challenge."

It is hoped that, by taking up the challenge of demonstrating that Elliot Smith made a significant and durable contribution to palaeopathology, this project will go some way towards ensuring that his legacy is finally recognised and fully acknowledged.

Acknowledgements

This work was supported by The Wellcome Trust [WT090575MA].

References

Crook, P., 2012. *Grafton Elliot Smith, Egyptology and the Diffusion of Culture. A Biographical Perspective.* Eastbourne (UK): Sussex Academic Press.

Dawson, W.R., ed., 1938. *Sir Grafton Elliot Smith: A Biographical Record by his Colleagues.* London: Jonathan Cape.

Granville, A.B., 1825. An essay on Egyptian mummies; with observations on the art of embalming among ancient Egyptians. *Philosophical Transactions of the Royal Society of London,* 1, pp.269-316.

Jones, F.W., 1938. In Egypt and Nubia. In W. Dawson, ed. 1938. *Sir Grafton Elliot Smith: A Biographical Record by his Colleagues.* London: Jonathan Cape, pp.139-148.

Lucas, A., 1962. *Ancient Egyptian Materials and Industries.* 4th ed., rev. J. R. Harris. London: Edward Arnold.

Murray, M.A., 1910. *The Tomb of Two Brothers.* Manchester: Sherratt and Hughes.

Osburn, W., 1828. *An Account of an Egyptian Mummy presented to the Museum of the Leeds Philosophical and Literary Society by the Late John Blayds, Esq.* Leeds: Leeds Philosophical and Literary Society.

Pettigrew, T.J., 1834. *A History of Egyptian Mummies.* London: Longman.

Reisner, G.A., 1910. *The Archaeological Survey of Nubia. Report for 1907-1908. Volume I. Archaeological Report.* Cairo: National Printing Department.

Ruffer, M.A., 1921. Histological studies of Egyptian mummies. In R.L. Moodie, ed., 1921. *Studies in the Palaeopathology of Egypt.* Chicago: University of Chicago Press.

Rutherford, L., 1938. Early days in Cambridge. In W. Dawson, ed. 1938. *Sir Grafton Elliot Smith: A Biographical Record by his Colleagues.* London: Jonathan Cape, pp.133-136.

Smith, G.E., 1902. On the Natural Preservation of the Brain in the Ancient Egyptians. *Journal of Anatomy and Physiology* 36 (4), pp.375-380.

Smith, G.E., 1912. *The Royal Mummies. Catalogue Général des Antiquités Égyptiennes de la Musée du Caire, Nos. 61051-61100.* Cairo: Service des Antiquités de l'Égypte.

Smith, G.E., 1938. Fragments of Autobiography. In W. Dawson, ed. 1938. *Sir Grafton Elliot Smith: A Biographical Record by his Colleagues.* London: Jonathan Cape, pp.113-120.

Smith, G.E. and Dawson, W.R. 1924. *Egyptian Mummies.* Reprinted 1991. London: Kegan Paul International.

Smith, G.E. and Jones, F.W., 1910. Report of the human remains. In G.A. Reisner, ed., 1910. *The Archaeological Survey of Nubia. Report for 1907-1908.* Cairo: National Printing Department, vol. 2, pp.7-367.

Wortham, J.D., 1971. *British Egyptology 1549-1906.* Newton Abbot (UK): David and Charles.

Whose body? The human remains from the 1908-1909 season of the Archaeological Survey of Nubia

Jenefer A. Cockitt

KNH Centre for Biomedical Egyptology, The University of Manchester, Manchester, UK

Abstract

The work of Sir Grafton Elliot Smith and his colleagues during the first Archaeological Survey of Nubia remains well known today due largely to the detailed nature of the anatomical report produced for the 1907-1908 season. Although the survey continued for a further three seasons, the achievements of the first year were not matched in terms of either productivity or reporting. Comprehensive anatomical or pathological reports were never produced for the bodies found in later years and many of those excavated have since become untraceable. This paper focuses on a recent attempt to locate the bodies from season two (1908-1909). Although it has been possible to put together a clearer picture of what was actually excavated and to identify a number of individual bones scattered across several collections, the provenance of much of the series appears to have been lost, along with the location of many of the bodies. The discovery of two sources of original documentary evidence for this collection has however made it possible to reconstruct the work carried out by the anatomists during this season. These records form a major resource for those working on the Archaeological Survey of Nubia and offer a unique opportunity to see the work of Elliot Smith and his colleagues completed for a further season.

Introduction

The Archaeological Survey of Nubia (1907-1911) remains well known in the field of bioarchaeology, despite having been carried out over a hundred years ago. This is due in part to the comprehensive anatomical report produced from the first season's work. Over 6000 bodies were excavated during this first season, which anatomists Sir Grafton Elliot Smith and Frederick Wood Jones studied to produce one of the earliest demographic surveys of the ancient Nubian population. Although not without its flaws, their report was extremely detailed for its time and recorded evidence of disease, trauma, possible medical treatment and artificial mummification, alongside a wide range of measurements intended to identify the height, sex and race of the skeletons found (Smith and Jones, 1910). The first season's work resulted in the production of not just an anatomical report, but also a corresponding archaeological report (written by the project director, George Reisner) (Reisner, 1910) and two short bulletins (Lyons et al., 1908; Reisner et al., 1908), written prior to the release of the annual reports and designed to highlight finds of interest to both anatomists and archaeologists. This was the pattern that was to be followed during the subsequent years of the survey – sadly, this did not happen.

Season two started on 1 October 1908 with much less enthusiasm than season one, and with several significant changes to the survey team. Frederick Wood Jones had left the survey, to be replaced by Douglas Derry. Both George Reisner and Grafton Elliot Smith were frequently absent from the excavations and by 1909 both had left – Elliot Smith for a Chair in Anatomy at The University of Manchester and Reisner for excavations in Palestine. Analysis of much of the material found during season one was still to be completed; although preliminary measurements were taken on some bodies, many had simply been collected and sent to Cairo for future study. The changes to the team, combined with the extremely large volume of material found during season one, meant that progress was slow during season two. The area covered (Kalabsha to Aman Duad) was not as productive as the first season and many bodies were found to have been disturbed and were therefore outside of their primary burial context. This was not ideal from the perspective of the anatomists who wanted intact, well dated bodies for their studies. Although Cecil Firth did go on to produce an archaeological report for season two (Firth, 1912), its anatomical partner was never produced.

The archaeological report alone is inadequate for anyone wanting to study the human remains found during this season. The report is very short and although it claims that the methodology established in the first season was followed (Firth, 1912, p.2) a number of apparent errors, particularly in the sequential numbering of graves and multiple burials, suggests that this was not the case. Despite this, it does appear that the anatomists had some input into the archaeological volume as the sex of many individuals is given and notes are included to highlight where especially large collections of human remains were made (such as cemetery 92). This cannot however be considered a combined report as many of the features of the first anatomical report are missing i.e. indicators of age, evidence of mummification or any reference to osteological measurements or evidence of disease.

The lack of an anatomical report has meant that the human remains from season two (and seasons three and four) have remained in relative obscurity. Although smaller and arguably less rewarding, the excavations were still extremely important for the study of ancient Nubian populations. The cemeteries (and therefore the human remains) concentrated upon in season one were those of the earliest periods of Nubian history, the A-group. The anatomists also examined skeletons and mummies from the Christian and Meroitic-Roman periods, due to the discovery of several very large Christian period cemeteries and Grafton Elliot Smith's own interest in mummification. Season two, in contrast, produced several large cemetery populations from the C-group and X-group (Ballana Culture) periods. As these were thus far under-represented in the Archaeological Survey of Nubia (ASN) collections, the study of human remains from these periods are important for a more complete picture demographically.

The only anatomical or pathological reports available for this season are two short bulletins, produced in 1909 (Reisner et al., 1909a; Reisner et al., 1909b). These detail a series of interesting examples of disease and trauma evident in the skeletal record. It is very clear from the tone of the bulletins however that the reports were only intended as preliminary notes. In several places, the reports consist of very detailed anatomical descriptions of a particular pathology, but with little accompanying discussion of a possible differential diagnosis. The anatomists intentionally only measured largely complete skeletons and Elliot Smith (1909, p.21) noted that from 2000 graves, only 300 were suitable for their studies. This does not however indicate that only 300 bodies were kept from these excavations. Grafton Elliot Smith, being a neuroanatomist, appears to have made a concerted effort to retain all skulls that were well preserved during season one and there is no reason to assume that a different practise was followed during season two.

The bulletins and the archaeological report together highlight the importance of the human remains found during season two both in terms of historically assessing the work of Grafton Elliot Smith and Douglas Derry and for future research into the ancient Nubian population. The following discussion therefore describes the human remains from season two that can currently be identified in collections around the world and the archival material that has been found relating to this part of the survey.

The surviving human remains

It has been possible to identify six collections to date that contain material from season two of the Archaeological Survey of Nubia: the KNH Centre for Biomedical Egyptology at The University of Manchester, The Natural History Museum, London, The Royal College of Surgeons, London, The Duckworth Laboratory at The University of Cambridge, the South Australia Museum, Adelaide and the Nubia Museum, Aswan. Although none hold a significant proportion of the human remains discovered, each collection plays an important role in reconstructing what happened to the bodies excavated during this season.

The KNH Centre for Biomedical Egyptology currently houses a small collection comprising a series of skeletal remains, together with five mummified heads, that once belonged to the late Professor of Anatomy in Manchester, Sir Grafton Elliot Smith. The skeletal material is thought to have once resided in the Anatomy Museum at the University (Warwick 1958, p.136), and after the closure of the museum, it was transferred to the anatomy department where it was used as a teaching collection for many years. Sadly, the provenance of much of the collection has now been lost, possibly as a result of high levels of handling over the years. Only 32 skulls and bones can now be positively identified, all but one of which come from the 1908-1909 season of the ASN (see table 1). The collection was once undoubtedly larger and although it probably contained material from other sites in Egypt worked upon by Grafton Elliot Smith, a significant amount of the collection was likely to have come from his time in Nubia. Two documents attest to the presence of additional ASN material in Manchester (Derry 1911, p.206; Watson, pre 1935); the latter records the presence of another 21 skulls from this series (see table 2), the location of which is no longer known.

The presence of material from this season in Manchester is not unexpected given that Elliot Smith was the lead anatomist on the project. He negotiated for the pathological material found during the survey to be sent to the Royal College of Surgeons in London (Smith, 1908), but this left hundreds or possibly thousands of bodies that showed no obvious evidence of disease or trauma. From the time he took up the Chair of Anatomy in Manchester, all of the human remains from the ASN that were to be sorted for distribution were routed through Elliot Smith in the UK. Any bones of pathological interest would be sent on to the Royal College of Surgeons (Smith, 1910b) or possibly even to Douglas Derry who helped sort many of the skeletons (Smith, 1910a) and was at that time based in University College London. Elliot Smith, for his part, was more interested in anatomical variation across the various time periods covered by Nubian history, through which he sought to answer questions about the racial history of the area (Smith, 1910d). There is no doubt that he built up his own collection from the human remains not required by the Royal College of Surgeons, but questions remain about the composition of that collection and its more recent history.

ASN number[1]	Sex	Period	Bone(s)
69:20	Female	C-group	Skull
72:239	Male	C-group	Skull
72:257	Male	C-group	Skull
72:365	Female	C-group	Radius
76:106	Female	C-group	Innominate
76:##	Unknown	Unknown	Radius
79:166	Male	A-group	Sacrum
87:33	Female	C-group	Skull
87:79	Male	C-group	Radius
87:89	Female	C-group	Skull
87:98	Male	C-group	Tibia, radius, ulna and both humerii
87:99	Male	C-group	Innominate, humerus
87:100	Male	C-group	Femur, both tibiae
87:104	Male	C-group	Both humerii, radius
87:117	Male	C-group	Sacrum, ulna
87:##	Unknown	Unknown	Femur
87:##	Unknown	Unknown	Humerus
87:##	Unknown	Unknown	Ulna
89:##	Unknown	Unknown	Fibula
89:##	Unknown	Unknown	Radius
89:##	Unknown	Unknown	Sacrum

Table 1: The provenanced bones from the ASN currently located in the KNH Centre for Biomedical Egyptology, The University of Manchester.

[1]The ASN number consists of the cemetery number, followed by the grave number. This conforms to the original numbers assigned to bodies during the survey.

ASN number	Sex	Period[1]
58: 106B	Male	C-group
69: 16	Male	C-group
72: 19B	Male	Christian
72: 19C	Male	Christian
72: 50	? Male	X-group
72: 79	Male	X-group
72: 88	Male	X-group
72: 165	Male / child	X-group
72: 241	Female	C-group
72: 392	Male	C-group
74: 68	Male	X-group or Christian
76: 35	Male	New Kingdom
77: 113	Male	A-group
87: 42	Female	A-group
87: 90	Male	A-group
87: 98	Male	A-group
87: 102	Female	A-group
87: 157	Female	A-group
91: 69	Female	A-group
92: 52	Male	Unknown
Number lost	Child	Unknown

Table 2: Skulls from the Archaeological Survey of Nubia that were originally in the Anatomy Museum at The University of Manchester (Watson, pre 1935).

[1]The dates for these burials have been taken from Firth (1912) or where no record was found, from the original record preserved in the Duckworth Laboratory archives, The University of Cambridge.

ASN number	Sex	Period	Bone(s)
58:##	Unknown	C-group	Vertebrae
58:##	Unknown	C-group	Vertebrae
72:41	Male	X-group	Skull
72:91	Female	X-group	Skull
89:686	Male	A-group	Left humerus, left scapula
89:754	Male	A-group	Left humerus
92:114	Male	X-group	Skull, mandible
92:121	Male	X-group	Skull

Table 3: The surviving human remains from the Nubian Pathological Collection that originated from the season two excavations.

The Natural History Museum, London today retains what remains of the Nubian Pathological Collection which once resided in the Royal College of Surgeons. Material was transferred to the Natural History Museum in the years after the Second World

War (Molleson, 1993). Sadly, the collection itself was badly damaged when the Royal College of Surgeons was bombed in 1941, resulting in the loss of significant proportion of the human remains, along with the card indexes providing provenance information on each sample. The Nubian Pathological Collection contains seven specimens from season two (see table 3) which includes several of the most important and unique examples of pathology or trauma found in the course of the entire survey. One specimen from the Nubian Pathological Collection – a cervical vertebra pierced by a bronze arrowhead – was retained by the Royal College of Surgeons where it remains on display to the general public. It is not known how much of the material from season two was lost when the RCS was damaged, but many of the examples of pathology and trauma noted in bulletins are not to be found in this collection or in any other. In the absence of additional archival evidence, it is not possible to reconstruct this part of the Nubian Pathological Collection.

The largest collection of material, both from the whole of the ASN and season two in particular, is housed in the Duckworth Laboratory at The University of Cambridge. The collection comprises bones from 23 bodies, mostly from cemeteries 71-76. These are mainly represented by a skull, with or without a mandible, although eight bodies are represented by post-cranial elements, three of which are substantially complete post-cranial skeletons. As with the material located in the KNH Centre, the majority of these bodies can be dated are from the C-group. Much of the season two material is in poor condition, with many of the skulls having suffered serious postmortem damage since excavation. There are however a number of interesting pathological examples in the collection including an ulcerated manubrium (72:368), an ossified larynx (74:15) and a benign osteoma (69:48).

Little is known about how this collection was formed; material from the ASN is recorded as residing in the Galton Laboratory at University College London prior to the outbreak of World War II (Batrawi, 1945, p.82). It is thought likely that this was transferred to Cambridge in the 1940s, possibly by Karl Pearson (Bellatti, personal communication, Dec 2011). If this is the case, it is possible that the collection was made by Douglas Derry during his time at University College London and possibly added to later by Sir Grafton Elliot Smith who became Chair of Anatomy there in 1919. The collection has received little attention to date and it remains possible that there is more material from season two to be found in the substantial uncatalogued part of the collection.

Only five artificial mummies from the Ptolemaic-Roman period cemeteries at Koshtamna (numbered 86 and 89) have been located to date. One resides in the South Australia Museum, Adelaide whilst the others were retained in Aswan and can now be found in the Nubia Museum. The records for the Koshtamna mummies are somewhat lacking, making it very difficult to evaluate the numbers found versus the numbers that might have been dissected by Smith and Derry in the course of the survey. The decoration of these mummies is highly variable, although some were found with cartonnage coffins and are in many ways indistinguishable from those found in Egyptian cemeteries. It therefore remains a possibility that museums may house one of the mummies from this site, but without a good provenance, it may not be possible to positively identify it.

ASN number	Sex	Period	Bone(s)
58:03	Male	C-group	Skull, mandible
58:14	Female	C-group	Skull
58:110	Male	C-group	Skull
58:119	Female	New Kingdom	Skull, mandible
60:##	Unknown	Unknown	Skull
62:##	Male	Unknown	Skull, mandible
69:04	Unknown	Unknown	Skull, mandible
69:48	Male	C-group	Skull
71:100	Unknown	Unknown	Skull, mandible
72:19A	Female	Christian	Coccyx
72:273	Female	C-group	Skull, mandible
72:368	Male	C-group	Sternum
72:445	Male	C-group	Skull, mandible
73:102	Male	C-group	Skull, mandible
73:108	Female	C-group	Skull, mandible
74:12	Male	Christian	Partial skeleton
74:15	Male	Christian	Skull, mandible, ossified larynx
75:88	Unknown	Unknown	Post-cranial skeleton
76:15D	Female	Unknown	Skull
76:81	Male	Unknown	Partial skeleton
76:87	Female	Unknown	Skull, mandible
83:107	Unknown	C-group	Post-cranial skeleton
87:39	Unknown	C-group	Skull, pelvis

Table 4: The provenanced bones from the ASN currently located in the Duckworth Laboratory, The University of Cambridge.

The original records for the 1908-1909 excavations

Following George Reisner's method of using recording cards for each grave, Elliot Smith devised a recording card to be used for each body found during the ASN (Smith, 1910c, p.10). In what is possibly the first example of its type, each card was printed to include sections for recording information such as age, sex, location, period, dentition, pathology and a wide range of measurements from the skull (see figure 1). These, along with field notebooks, formed the basis of the bulletins and the first anatomical report. The cards were brought back to England from Egypt on the completion of the survey and were last recorded as being made available to those studying the ASN material during the 1940s (Batrawi, 1945). It had been assumed that the recording cards had since been lost, until the author rediscovered 495 of them in the Duckworth Laboratory archive at The University of Cambridge in 2011. Fourteen of these belong to cemeteries 50 and 55 from season one and two are illegible - the remaining 479 come from season two.

The cards, which are often difficult to read, contain details of 750 bodies from 18 cemeteries. A number of cards contain records for multiple bodies, particularly where

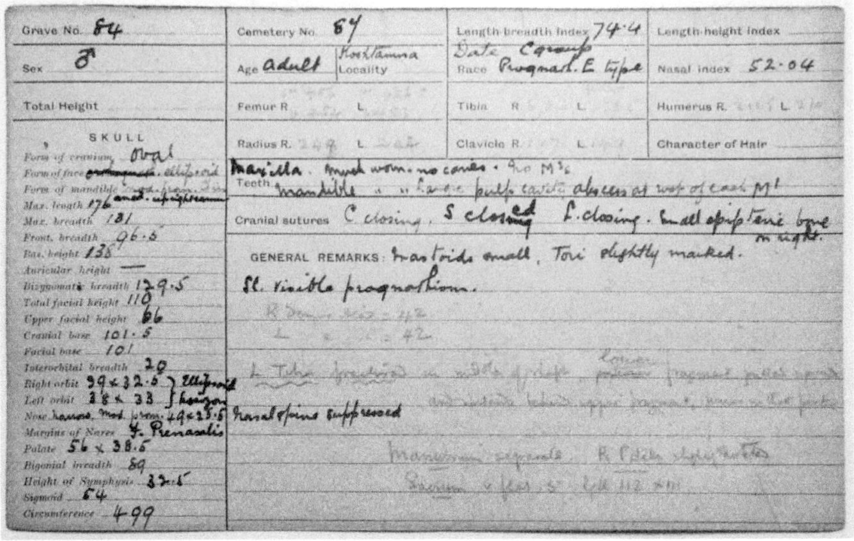

Figure 1: An example of a surviving ASN tomb card from cemetery 87, grave 84. Photography courtesy of the Duckworth Laboratory, The University of Cambridge.

a number of infants, children or poorly preserved bodies were found in one area. The information provided for each body is, not surprisingly, highly variable and is usually a reflection of the degree of preservation found. These cannot be considered a complete set of cards for season two as there are no cards for a number of cemeteries (60, 61, 63, 64, 65, 66, 67, 70, 75, 80, 81, 82, 83, 84, 90, 91, 92 of which cemeteries 60, 61, 63 and 84 are also not included in the catalogue of graves in the archaeological report).

The approach of the authors of the cards (Grafton Elliot Smith, Douglas Derry and a third, possibly H.W. Beckett) in regard to the description of observed trauma or pathology is interesting. Fractures are reasonably well reported, with indicators of the degree of union and shortening given, alongside notes on the presence or absence of any infection. These are not as consistently well reported as those found in the first season's report, however, it must be remembered that the cards represent mostly the initial findings of the anatomists and not necessarily their final, measured opinion. Some of the pathologies they observed are described in detail, but these are mostly restricted to those affecting the skull. Some of the more unusual examples were identified as needing to be described at a later date. In most cases where this occurs, the pathology was described often in superb detail in one of the two bulletins for the season, demonstrating that the anatomists did revisit the skeletal material from this season and were able to carry out some further analysis. Additional evidence of this is provided with the length-breadth and nasal indexes for the skull which were added in pen to the cards, presumably at a later date, and by the small number of revisions or ticks of agreement added to long bone measurements.

The cards provide the answers to several questions raised in the course of studying any of the ASN reports or human remains. More than 7500 bodies are described in the reports, but traces can be found for around 2000 of them, begging the question – what happened to the rest? Comments on the cards about the retention of specific bodies appear to indicate that not all bodies were kept, especially those in a fragmentary condition or those without a head. Queries that the anatomists had over the age and/or sex of a body at the time of excavation were not conveyed to the archaeological report. It may be that these issues were revisited and resolved or that the issues were of minor importance to Firth who simply removed the query. The problems with the aging and sexing of skeletons in the early 20th century are well known; an explanation for some of the problems may be found here.

Many of the surviving human remains from the ASN are difficult to match to their original grave, especially where the grave contained multiple burials. It is usually assumed that this is due to the passage of time and high levels of handling. The cards however have identified this as an issue at the time, which makes most problems now impossible to solve. There are discrepancies between the anatomical and archaeological reports about whether some graves contained bodies and in some cases, the age of the body (such as adult versus infant). The system of using capital letters to identify multiple bodies in a single tomb, used in season one, appears to break down quite often. In some instances where the second body found in a grave was an infant or child a character was not assigned. The archaeological report often omits the letters assigned to bodies as they are recorded on the cards and in a small number of cases a character was added to a grave rather than a body. The cards allow a unique opportunity to correct some of the errors identified on both sides, and highlight the cause of problems encountered with material from the other seasons.

The George Reisner archives, Boston

The Museum of Fine Arts, Boston is currently the major repository for archival material belonging to the archaeologist, George Reisner. As well as containing a huge archival series from his work at Giza, the Museum of Fine Arts also holds a sizeable collection of glass lantern slides and photographic prints from the Archaeological Survey of Nubia. The photographs belong mainly to seasons one and two, the years when Reisner himself directed the excavations. Although a significant number were published in the annual reports, the archive also holds a collection of unpublished photographs. Those from season two are now especially valuable, given the absence of an anatomical report and the subsequent loss of most human remains from those excavations.

The archival photographs suggest an anatomical report for season two was originally intended; a number of both anterior and lateral images of skulls, in the same style as those in both the season one report and several of the bulletins, are preserved (for example figure 2). None of those shown are known to survive, making these photographs the only visual record of those skulls. The skulls shown come mostly from the better preserved cemeteries (74, 76, 77, 79 and 87) and are usually a pairing of one male and one female skull. A large number of photographs of the mummies found at Koshtamna also reside in the archive. These directly reflect the mummies included in the archaeological report, although a number of images showing very slight variation

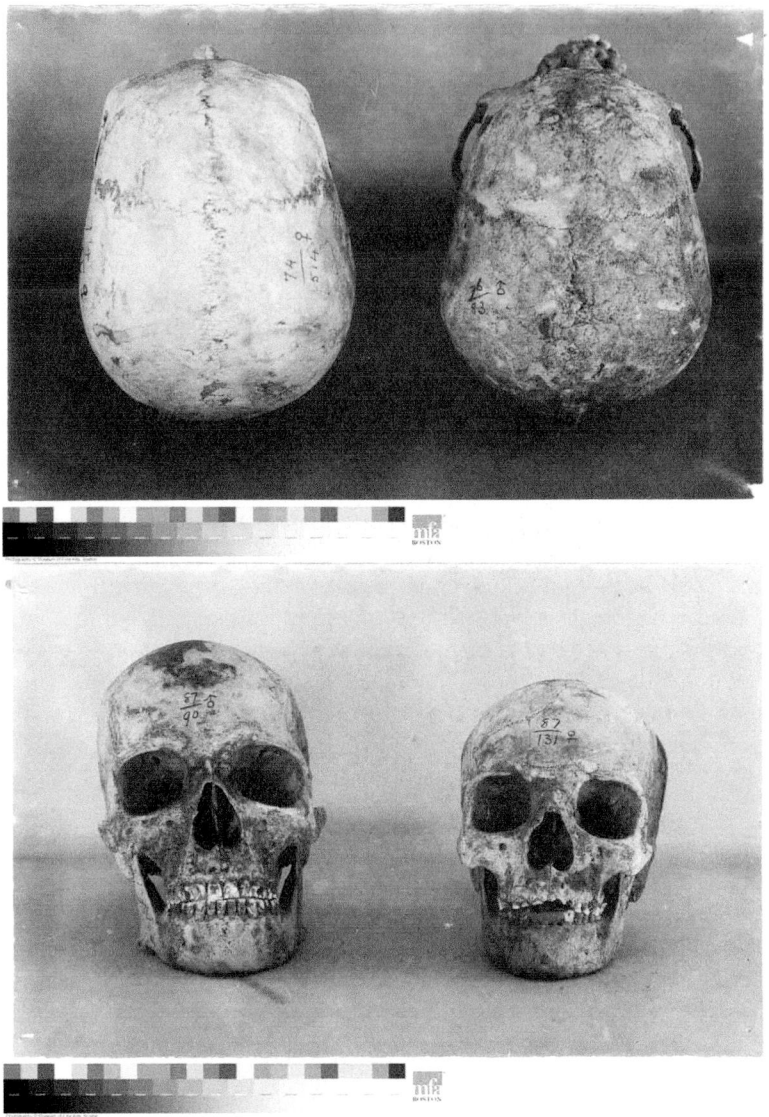

Figure 2: Top – The top of two skulls from 74/514 and 76/83 ; Bottom – The faces of two skulls from 87/90 and 87/131
1909 Archaeological Survey of Nubia. Courtesy Museum of Fine Arts, Boston. Photographs © [2014] Museum of Fine Arts, Boston

from the published versions are also included. Interestingly, there are no images of the mummies Smith and Derry dissected and described on report cards and likewise there are no cards for the mummies for which photographs were taken. Those photographed were largely intact and in some cases were elaborately wrapped; it seems the rule of not dissecting intact mummies established in season one (Smith, 1910e, p.66) persisted into season two at least.

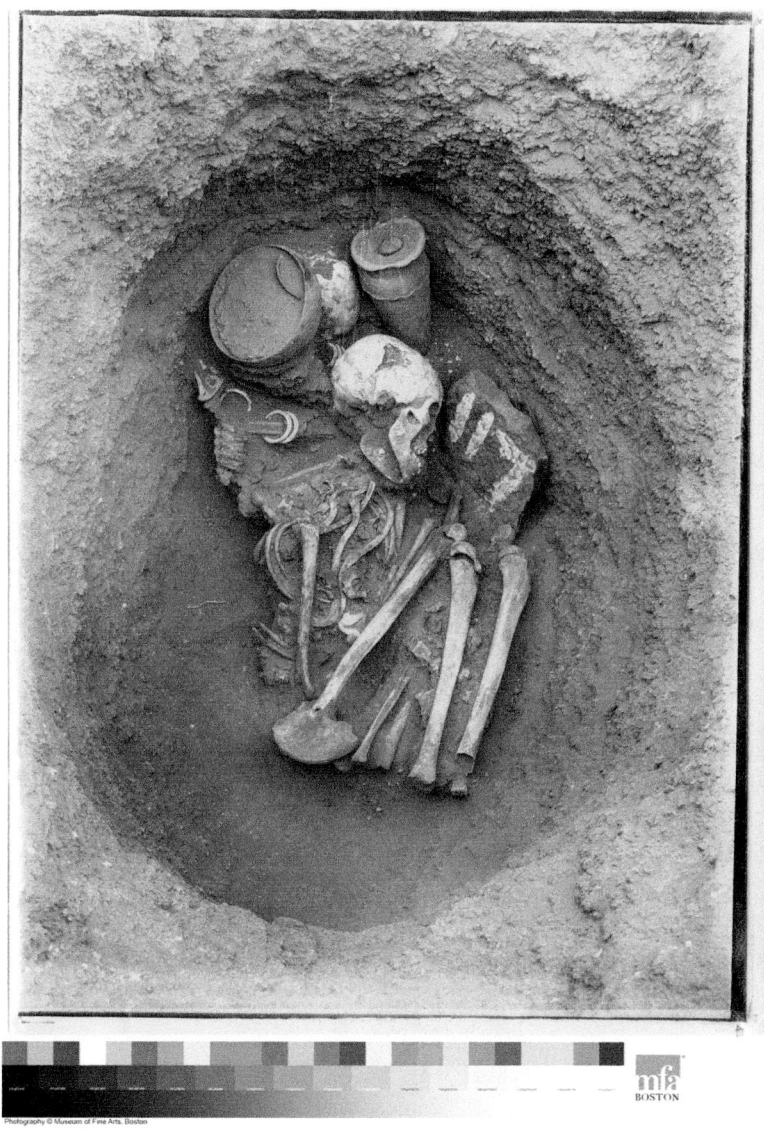

Figure 3: Photograph B763 excavated skeleton in grave
– cemetery 79, grave 117, 1909 Archaeological Survey of Nubia.
Courtesy Museum of Fine Arts, Boston. Photograph © [2014] Museum of Fine Arts, Boston

A considerable number of photographs were retained from cemetery 69 by Reisner. These feature prominently in the archive, but rarely in the archaeological report. The majority of these are images of graves post excavation and show in excellent clarity the position of the body (or bodies), the grave goods buried with them and provide an indication of any major skeletal elements that may be absent (such as figure 3). A similar situation is found with cemetery 80, although there are fewer examples. Derry

(1909, p.22) notes that the burials in this latter cemetery were found to be too fragile to handle after excavation, so Reisner made a photographic record of them. It is possible this was also the case for some of the burials in cemetery 69. No bodies to date have been found from cemetery 80, whilst just four have come to light from cemetery 69. The photographs taken by Reisner appear to have been important to the anatomists as a record of the presence of bodies in these cemeteries, but of limited interest to the archaeologists who omitted them from the season two report.

A wide range of panoramic shots of cemeteries 82, 85 and 86 were taken at various stages of excavation and were not selected for publication, but were retained in the archives. As well as providing an amazing record of cemeteries that no longer exist, the photographs also provide insight into the landscape that prevailed in the area and the general preservation conditions. Preservation appears to have been much poorer in this area than that experienced during season one. This seems to have been due to wide range of factors including significant damage caused by tomb robbers and sebbakh diggers (Reisner, 1909, p.17), whose activities are well recorded photographically. The considerable depth of C-group burials is also well illustrated in the photographic records, and suggests another potential reason for poor preservation.

Conclusions

Although considerable progress has been made towards reconstructing the work of Sir Grafton Elliot Smith and Douglas Derry during season two of the ASN, it now appears unlikely that a comprehensive collection of the human remains from these cemeteries will be discovered. The human remains that do survive are few in number and are distributed throughout museum collections in the UK, Egypt and Australia. The material retained from this season was limited, both in the number of actual bodies and the range of cemeteries they were from, due largely to the poor condition of finds in many areas. The archival sources indicate strongly that the main body of material from this season was brought to and kept in Manchester, forming the backbone of Elliot Smith's anthropological collection. It is therefore also apparent that it is from this collection that many bodies have been damaged, lost or transferred without record.

Despite this, the surviving archival records have offered a way of completing the work begun by Sir Grafton Elliot Smith and Douglas Derry over a hundred years ago. A partial anatomical report can now be produced for the 1908-1909 season, adding significantly to the examples of pathology and trauma recorded for this collection and offering a resource for those continuing to work on the Archaeological Survey of Nubia in the future.

Acknowledgements

This work was supported by The Wellcome Trust [WT090575MA].

Grateful thanks are due to a large number of people including Rob Kruszynski at The Natural History Museum, London and Marta Lahr and Maggie Bellatti at the Duckworth

Laboratory, The University of Cambridge for all of their help with access to the ASN material; Denise Doxey and Rita Freed at the Museum of Fine Arts, Boston for their help with accessing George Reisner's ASN archives; Ryan Metcalfe for taking and editing a huge number of photographs and Rachel Harding for her help cataloguing the KNH Collection.

References

Batrawi, A., 1945. The Racial History of Egypt and Nubia. *The Journal of the Royal Anthropological Institute of Great Britain and Ireland* 75 (1-2), pp. 81-101.

Derry, D.E., 1909. Field Notes. In: C.M. Firth, G.A. Reisner, G.E. Smith and D.E. Derry, eds. *The Archaeological Survey of Nubia. Bulletin IV.* Cairo: National Printing Department, pp.22-28.

Derry, D., 1911. Note on Accessory Articular Facets between the Sacrum and Ilium, and their Significance. *Journal of Anatomy and Physiology* 45 (3), pp.202-10.

Firth, C.M., 1912. *The Archaeological Survey of Nubia. Report for 1908-1909.* Cairo: Government Press.

Lyons, H.G., Reisner, G.A., Smith, G.E. and Jones, F.W., 1908. *The Archaeological Survey of Nubia. Bulletin I.* Cairo: National Printing Department.

Molleson, T.I., 1993. The Nubian Pathological Collection in the Natural History Museum, London. In: W.V. Davies and R. Walker, eds. *Biological Anthropology and the Study of Ancient Egypt.* London: British Museum Press, pp.136-43.

Reisner, G.A., Smith, G.E. and Jones, F.W., 1908. *The Archaeological Survey of Nubia. Bulletin II.* Cairo: National Printing Department.

Reisner, G.A., 1909. The Archaeological Survey of Nubia. In: G.A. Reisner, G.E. Smith and D.E. Derry, eds. *The Archaeological Survey of Nubia. Bulletin III.* Cairo: National Printing Department, pp.5-20.

Reisner, G.A., Smith, G.E. and Derry, D.E., 1909a. *The Archaeological Survey of Nubia. Bulletin III.* Cairo: National Printing Department.

Reisner, G.A., Firth, C.M., Smith, G.E. and Derry, D.E., 1909b. *The Archaeological Survey of Nubia. Bulletin IV.* Cairo: National Printing Department.

Reisner, G.A., 1910. *The Archaeological Survey of Nubia. Report for 1907-1908. Volume I. Archaeological Report.* Cairo: National Printing Department.

Smith, G.E., 1908. *Letter to Sir Arthur Keith.* Manchester 24th April. RCS-MUS/5/3/3. Royal College of Surgeons Archive, London.

Smith, G.E., 1909. Anatomical Report A. In: G.A. Reisner, G.E. Smith, D.E. and Derry, eds. *The Archaeological Survey of Nubia. Bulletin III.* Cairo: National Printing Department pp.21-27.

Smith, G.E., 1910a. *Letter to Sir Arthur Keith.* Manchester 2nd May. RCS-MUS/5/3/3. Royal College of Surgeons Archive, London.

Smith, G.E., 1910b. *Letter to Sir Arthur Keith.* Manchester 28th June. RCS-MUS/5/3/3. Royal College of Surgeons Archive, London.

Smith, G.E., 1910c. Introduction. In: G.E. Smith and F.W. Jones, eds. *The Archaeological Survey of Nubia. Report for 1907-1908. Volume II. Report on the Human Remains.* Cairo: National Printing Dept, pp.7-14.

Smith, G.E., 1910d. The Racial Problem. In: G.E. Smith and F.W. Jones, eds. *The Archaeological Survey of Nubia. Report for 1907-1908. Volume II. Report on the Human Remains*. Cairo: National Printing Dept, pp.15-36.

Smith, G.E., 1910e. An account of the fieldwork in the neighbourhood of Shellal. In: G.E. Smith and F.W. Jones, eds. *The Archaeological Survey of Nubia. Report for 1907-1908. Volume II. Report on the Human Remains*. Cairo: National Printing Dept, pp.37-106.

Smith, G.E. and Jones, F.W., 1910. *The Archaeological Survey of Nubia. Report for 1907-1908. Volume II. Report on the Human Remains*. Cairo: National Printing Dept.

Warwick, R., 1958. Frederick Wood Jones and the Spirit of Anatomy. *Manchester University Medical School Gazette* 37, pp.135-44.

Watson, J.V., pre 1935. *Osteometrical Data* [Workbook]. RCS-MUS/7/8/13. Royal College of Surgeons Archive, London.

The more things change?
The archaeological work of Alfred Lucas

Ryan Metcalfe

KNH Centre for Biomedical Egyptology, The University of Manchester, Manchester, UK

Abstract

Although Alfred Lucas was not the first person to analyse ancient Egyptian material, he is certainly among the most famous, not least because his seminal work on the subject was one of the only such texts available for many years. His work covered a wide range of material analysis, from the composition of copper objects to the nature of cosmetics. Of great importance to Egyptologists with an interest in mummification is his painstaking research into the use of natron in mummification which, through a series of carefully performed experiments supported by a number of additional sources, finally laid to rest the question of how this desiccant was used. As many of the modern analytical tools available to archaeological science were not available to Lucas, questions may be raised about the applicability of his work in the modern day. This paper will attempt to demonstrate the relevance of his work to modern biomedical Egyptology.

Introduction

As with many of his contemporaries, Alfred Lucas took up residence in Egypt because of ill health, in his case tuberculosis (Balls et al., 1946). Once recovered, in 1898, he joined the Salt Department as a chemist, continuing a career in the subject that began with a position in the Laboratory of the Government Chemist in London several years earlier (Gilberg, 1997). It is really from this on point that his work starts to interest the Egyptologist, but his tenure in Egypt is of surprising relevance to a number of other academic disciplines including forensic science. His text book on this subject, 'Forensic Chemistry', may be surprisingly short on experimental detail, but it is wide in scope, covering many subjects of interest to forensic scientists such as explosive and firearm analysis, methods for detecting counterfeit coins, and what would today be referred to as good laboratory practice. The chapters include illustrative examples from a wide range of interesting cases including some with a basis in archaeology – the section on preservation of the human body contains a great deal of information gained by his examination of Egyptian and Nubian mummies, for example (Lucas, 1921, pp.221-234).

His experience in so many areas of chemical and forensic analysis proved to be of great relevance to the study of antiquities, and this may be what he is best remembered for today. There are many crossovers between forensic and archaeological science, with many of the same questions being asked. Possibly the most common of these being simply, what is this item made of? Lucas used his skills to investigate a large number of different artefacts including residues of cosmetics and perfumes (Lucas, 1930a), embalming materials (Lucas, 1931), pottery (Lucas, 1929) and faience (Lucas, 1936).

The crowning glory of this career in archaeological research came in the form of 'ancient Egyptian Materials and Industries' (Lucas and Harris, 1962), a book that has seen several editions and is still available and still relevant nearly 90 years after first publication. The next comparable work did not appear until 2000 (Nicholson and Shaw, 2000), and required the expertise of more than thirty contributors rather than being the work of a single man. Indeed, so complete and well regarded was Lucas' book that it may even have had a negative impact on the development of 'Egyptological science'. In a discussion of the relationship between Egyptology and the sciences, Nicholson states that "For most Egyptologists the science had been *done* by Lucas and he had provided all that it was necessary to know" (Nicholson, 2010, p.124, emphasis as in source).

However, despite the vast amount of work he carried out, and the great number of papers and other works in which it was reported, it can be difficult to determine exactly how Lucas conducted his analyses. Occasional hints are provided, such as a mention that samples were combusted to determine the organic content, or that material was sent to an expert for spectrometric analysis, but details of the equipment used or protocols employed are scarce. For a practicing scientist used to the modern reporting methods required by journals, this can be hugely frustrating though it should perhaps not be too surprising. Throughout Lucas' publications, and in those of his contemporaries, methods used for chemical identification are referred to simply by name. It would appear, therefore, that the use of standard texts (such as Scott, 1917) was normal, and that scientists would be able to recognise them or be able to reference them readily enough that further description was unnecessary.

State of his art

Many of the methods that a modern archaeological scientist would consider to be standard were in the early stages of development, theoretical, or completely unknown during Lucas' working life. Chromatography, which is commonly used today to separate out complex mixtures of chemicals in, for example, embalming materials or pot residues, may have a very long history. Both Moses and Pliny the Elder have been suggested as early proponents of chromatography, though there is some doubt that this claim bears up to scrutiny (Ettre, 1995). Simple methods based on paper chromatography became a common tool in the late 1800s. One of the most commonly used forms of the technique in archaeological research, gas chromatography (GC), was in its infancy at the end of Lucas' life and only really started to reach maturity in the 1950s and 60s (Touchstone, 1993). Mass spectrometry, a common partner to GC as it helps to identify the compounds that have been separated, has a somewhat shorter history. The first developments were made around the turn of the 20th century, but it was used almost solely by physicists for the first few decades of its existence. It only started to become popular among organic chemists at about the same time as GC (Griffiths, 2008). The genetic methods that fire the public's imagination so intensely these days are significantly more modern than either of these methods. The structure of DNA was only determined in the 1950s (Watson and Crick, 1953), with the amplification methods needed for the study of ancient DNA developed in the mid-1980s (Saiki et al., 1985).

The standard methods available during Lucas' Egyptological work would have been based significantly on a combination of physical properties and chemical reaction. Scott's Standard Methods of Chemical Analysis, for example, suggests that the initial examination of an unknown oil can include looking for fluorescence, testing the odour and even tasting it before moving on to other tests including the determination of the flash point (Scott, 1917, pp.566-568). Methods for identifying specific chemicals were frequently based on converting the substance in question into a specific chemical and purifying it by reaction before calculating the original amount by weight or titration.

In addition to his own skills, those of others would be put to use as required. This was not limited to areas outside of his expertise, such as the use of an entomologist to identify insect parts in what was thought to be bat faeces found within skulls from ancient Nubia (Lucas, 1910), but also included instances where the limitations of fieldwork came into play. Such was the case during the excavation and field-conservation of the artefacts from the tomb of Tutankhamun, where Lucas limited himself to simple analyses that were required for material analysis, with several other samples sent to chemists for more thorough scrutiny (Lucas, 1963).

Reliability

Given that the methods used by Lucas and his collaborators may differ greatly from the tools that we would use today, are his data still useable and comparable to those collected by more modern researchers? Due to the great range of subjects, this is not an easy question to attempt to answer.

Many of his papers do not appear to have used analytical chemistry to any great extent. His work on copper in ancient Egypt, for example, is primarily a review of the archaeological evidence for copper mining, such as inscriptions, the presence and location of ancient slag heaps, and physical evidence of copper items or copper ores in graves (Lucas, 1927). A paper on Predynastic stone vessels similarly revolves around the identification of where deposits of specific stones may be found. The small amount of analysis underpinning this is based on microscopic examination of small pieces from broken artefacts to identify the stone from which they were made, with confirmation provided by geologists (Lucas, 1930b). In cases such as these, the simple methods involved make the data difficult to dispute barring a significant change in understanding.

Several other papers have a strong basis in experiment to test or reproduce ancient processes. These include a paper on the nature of black and red colouration in pottery, of special interest to ancient Egypt thanks to the production of black-rimmed ware in the Predynastic (Lucas, 1929). This treatise includes several examples of experiments either on clay or on pieces of ancient pottery to demonstrate under what circumstances these colours may be interchangeable and how two-coloured pots may be made in a single firing without the use of coloured slips or coatings. The simplicity of these experiments, based necessarily on simple heating and different firing environments (e.g. exposure to smoke), are very difficult to refute. Indeed, the paper itself is highly critical of more complex experiments that produce misleading results precisely because of this complexity. For example, the production of a black colour by conversion of ferric oxide to ferrous oxide in a hydrogen atmosphere at lower temperatures than are sufficient to fire pottery is an unlikely mechanism for ancient potters to have used.

Possibly the most famous example of this type of experimental work, at least for those with an interest in mummification, was based on the use of natron for the preservation of human and animal remains (Lucas, 1932). This subject had been the source of some debate, with opinion divided as to whether bodies were immersed in a bath of natron or covered with the dry salt. The paper as a whole is an excellent example of the carefully considered, multi-disciplinary approach that should be taken when investigating ancient technologies. The physical evidence from material finds and mummies themselves, ethnographic evidence of similar methods, linguistic evidence from Herodotus' account, and data from experimentally produced mummies are all presented and assessed in turn, with summaries provided along the way to help marshal the disparate sources employed. The conclusions would have been far less convincing if the models had been presented by themselves. The pigeons and chickens used are far smaller than a human corpse, for example, so the rate of desiccation would likely be far greater in the experiment. More recent experiments on isolated human body parts have shown that the process of mummification may be more complex or more sensitive to environmental conditions than previously thought (Papageorgopoulou et al., 2011), so extrapolation from such small body models to the human scale may not be easy. However, the volume and range of additional evidence that Lucas provides gives great weight to his conclusions, and to his determination that natron was used in the dry form.

The nature of these studies means that they can still be used as reliable sources of information today, despite the radical advances in technology that the intervening decades have provided to archaeology. What must be used more cautiously, however, are data obtained through the chemical analysis of ancient organic materials. Today, the analysis of organic residues is usually based on sophisticated tools such as coupled chromatography (gas or liquid) and mass spectrometry to separate and identify the constituents of a complex material, spectroscopic methods for chemical bond identification and a host of other methods (for a review of the application of these to ancient cosmetics and pharmaceuticals, for example, see Ribechini et al., (2011)). In addition to being minimally or non-destructive, these methods are extremely sensitive. By comparison, the methods that Lucas would have employed would have had far higher material requirements and minor components may well have been missed. These minor components can be extremely useful, for example squalene may be used as in indicator of modern contamination in lipid profiles (Evershed, 1993). In terms of the amount of material required for analysis, standard methods in the early 20th century would have required more than half a gram of solid fat just for testing the amount of unsaturated lipid in a sample (Scott, 1917, p.584). Chromatographic analysis of archaeological materials require far less, on the order of a few grams from materials such as pottery or bone, and around 1000 times less for dried tissues (Evershed et al., 2002). It is possible that such small samples may produce misleading results if the specimen is inhomogeneous, but this concern is more than outweighed by the significantly smaller amount of damage inflicted upon the specimen in taking those samples.

Hidden information

With such a vast array of material to his name, it is perhaps not surprising that some of the data Lucas provides has gone unremarked of late. Some of this has become particularly relevant and in need of retesting as new avenues of research have opened up. The use of ancient DNA analysis has been of great interest in recent years given its power and applicability to a wide range of research areas. However, the application of this tool to ancient Egyptian artefacts has been dogged by controversy (see Marchant, 2011 for a review). One of the supporting pieces of evidence in favour of DNA preservation in Egypt has been that natron, being an alkaline salt on account of the high levels of sodium carbonate and sodium bicarbonate it contains, should help to protect DNA from degradation. Statements such as "Therefore, the use of natron creates a neutral or mildly alkaline environment that significantly enhances DNA preservation" (Zink and Nerlich, 2003) and "An alkaline environment, such as the one provided by mummification in alkaline natron, is also beneficial to the survival of the acidic DNA molecule" (Rutherford, 2008, p.121) can be found in the literature, though they are not usually supported with a reference or with empirical data.

Whilst it is certainly plausible that the use of a strongly alkaline substance such as natron would raise the pH of soft tissues as it desiccates them, this may not always be true. Lucas (1932, p.137) states that:

> "One argument that has been used against the employment of natron in embalming is that mummies are generally, though not invariably, acid and not alkaline; yet a body may have been treated with natron and still be acid, as is proved by the two pigeons mummified by the writer"

The explanation for this apparently paradoxical observation is that the increase in pH caused by the use of natron is overwhelmed by the presence of fatty acids and acidic decomposition products. This in itself may sound counter-intuitive, as the purpose of mummification is to prevent decomposition taking place. However, as mentioned above, recent research into mummification has shown that the process may be more poorly understood than commonly assumed, and that simple burial in natron may not be sufficient to stop decomposition (Papageorgopoulou et al., 2011). Given the potential impact that this simple property of Egyptian mummies may have, further research may be required. This is especially true given that the use of natron may be highly variable across both history and social class, which may in turn result in substantial variation in DNA preservation from this simple factor alone.

Conclusion

The sheer volume and scope of Lucas' archaeological work is truly impressive, but perhaps even more so is that so much of it remains valid and useable up to a century later. This is due partly to the amount of modelling and reproduction that his research used, and partly thanks to the care and diligence with which he approached analysis. Although the march of time has brought about developments in technology and changes to our understanding of ancient Egypt, it is likely that Lucas will remain a valuable source for many years to come.

References

Balls, W.L., Engelbach, R., Gracie, D.S., Hurst, H.E. and McCallum, L.F., 1946. Mr. Alfred Lucas, O.B.E. *Nature*, 157 (3988), p.433.

Ettre, L.S., 1995. Chromatography in the Ancient World? *Journal of High Resolution Chromatography*, 18, pp.277-278.

Evershed, R.P., 1993. Biomolecular Archaeology and Lipids. *World Archaeology*, 25 (1), pp.74-93.

Evershed, R.P., Dudd, S.N., Copley, M.S., Berstan, R., Stott, A.W., Mottram, H., Buckley, S.A. and Crossman, Z., 2002. Chemistry of Archaeological Animal Fats. *Accounts of Chemical Res*earch, 35, pp.660-668.

Gilberg, M., 1997. Alfred Lucas: Egypt's Sherlock Holmes. *Journal of the American Institute of Conservation*, 36 (1), pp.31-48.

Griffiths, J., 2008. A Brief History of Mass Spectrometry. *Analytical Chemistry*, 80 (15), pp.5678-5683.

Lucas, A., 1910. Chemical Report on Samples of Various Materials Found in Nubia During the Progress of the Archaeological Survey. In: G. E. Smith and F. W. Jones, *The Archaeological Survey of Nubia Report for 1907-1908, Volume II: Report on the Human Remains*. Cairo: National Printing Department, pp.371-374.

Lucas, A., 1921. *Forensic Chemistry*. London: Edward Arnold & Co.

Lucas, A., 1927. Copper in Ancient Egypt. *Journal of Egyptian Archaeology*, 13 (3/4), pp.162-170.

Lucas, A., 1929. The Nature of the Colour of Pottery, with Special Reference to that of Ancient Egypt. *Journal of the Royal Anthropological Institute of Great Britain and Ireland,* 59, pp.113-129.

Lucas, A., 1930a Cosmetics, Perfumes and Incense in Ancient Egypt. *Journal of Egyptian Archaeology*, 16 (1/2), pp.41-53.

Lucas, A., 1930b. Egyptian Predynastic Stone Vessels. *Journal of Egyptian Archaeology* 16 (3/4), pp.200-212.

Lucas, A., 1931. 'Cedar'-Tree Products Employed in Mummification. *Journal of Egyptian Archaeology*, 17 (1/2), pp.13-21.

Lucas, A., 1932. The Use of Natron in Mummification. *Journal of Egyptian Archaeology*, 18 (3/4), pp.125-140.

Lucas, A., 1936. Glazed Ware in Egypt, India, and Mesopotamia. *Journal of Egyptian Archaeology*, 22 (2), pp.141-164.

Lucas, A. and Harris, J.R., 1962. *Ancient Egyptian Materials and Industries*. 4th ed. London: Edward Arnold (Publishers) Ltd.

Lucas, A., 1963 Appendix 2: The Chemistry of the Tomb. In: H. Carter, *The Tomb of Tut•Ankh•Amen*. New York: Cooper Square Publishers Inc., pp.162-188.

Marchant, J., 2011. Ancient DNA: Curse of the Pharaoh's DNA. *Nature*, 472, pp.404-406.

Nicholson, P.T. and Shaw, I., 2000. *Ancient Egyptian Materials and Technology*. Cambridge: Cambridge University Press.

Nicholson, P.T., 2010. 'Other Than' – Egyptology as Science? A Selective History. In: J. Cockitt and R. David, eds. *Pharmacy and Medicine in Ancient Egypt – Proceedings of the conferences held in Cairo (2007) and Manchester (2008)*. Oxford: Archaeopress, pp.122-126.

Papageorgopoulou, C., Shved, N. and Ruhli, F.J., 2011. Post-mortem Alterations of Mummified Human Tissue Under Experimental Setting (abstract). In: 7th World Congress on Mummy Studies Program with Abstracts, San Diego, CA. p.90.

Ribechini, E., Modugno, F., Perez-Arantegui, J. and Colombini, M.P., 2011. Discovering the Composition of Ancient Cosmetics and Remedies: Analytical Techniques and Materials. *Analytical and Bioanalytical Chemistry*, 401, pp.1727-1738.

Rutherford, P., 2008. DNA Identification in Mummies and Associated Material. In: R. David, ed. *Egyptian Mummies and Modern Science*. Cambridge: Cambridge University Press, pp.116-132.

Saiki, R.K., Scharf, S., Faloona, F., Mullis, K.B., Horn, G.T., Erlich, H.A. and Arnheim, N., 1985. Enzymatic Amplification of β-Globin Genomic Sequences and Restriction Site Analysis for Diagnosis of Sickle Cell Anemia. *Science*, 230 (4732), pp.1350-1354.

Scott, W.W., 1917. *Standard Methods of Chemical Analysis*. New York: D. Van Nostrand Company.

Touchstone, J.C., 1993. History of Chromatography. *Journal of Liquid Chromatography*, 16 (8), pp.1647-1665.

Watson, J.D. and Crick, F.H.C., 1953. A Structure for Deoxyribose Nucleic Acid. *Nature*, 171 (4356), pp.737-738.

Zink, A. and Nerlich, A.G., 2003. Molecular Analyses of the 'Pharaos:' Feasibility of Molecular Studies in Ancient Egyptian Material. *American Journal of Physical Anthropology*, 121, pp.109-111.

Harris lines, ill health during childhood, poor diet, emotional stress or normal growth patterns?

Abeer Eladany

University of Aberdeen, Aberdeen, UK

Abstract

This paper will discuss Harris lines which is one of the most common observations reported from ancient Egyptian mummies. Harris lines are transverse lines of higher density or radio-opaque lines in the metaphyses that can be seen in the radiographs of long bones (usually tibia, femur and radius). Enamel hypoplasia are similar lines seen across the enamel of the teeth and were regarded to be associated with Harris lines.

Although frequently reported, physical anthropologists, pathologists and biomedical Egyptologists still do not agree on a specific explanation or a definite cause of the lines. Periods of ill health, poor diet and emotional stress were suggested as strong candidates but no research has been conducted which provided simple and clear answers. Some of the results of a recent radiological study of a collection of seven mummies from the British Museum will be presented here.

Harris lines: an overview

Harris lines were named after H. A. Harris, who reported them in 1931 when he radiologically examined a large number of skeletons of still born children. He associated the lines with periods of malnutrition of the foetus as the result of an illness that affects the pregnant mother (Harris, 1931a, p.622).

However, this was not the first mention of the lines in scientific literature. The lines were reported in 1926 by Harris himself and even much earlier, in 1874 by Wegner who observed the lines while examining cross sections of bones of rabbits and chickens, according to Wells (1963, p.406) who described them as a new approach to palaeopathology.

In 1926, Elliot Smith mentioned the lines reported by Harris:

> "He has been able to demonstrate by anatomical studies that any diminution of the growth processes of the body (such as normally occur during the week after birth and in certain phases of puberty, as well as in grave illness, or even deprivation of a diabetic patient of his dose of insulin), will cause an opaque line to make its appearance in X ray photographs of bones. This has been mistaken for, and used as, a diagnostic sign of healed rickets."
>
> (Smith, 1926, p.946)

Harris published his research while working as an anatomist when Elliot Smith was the Chair of Anatomy at University College London. In 1951, Harris wrote:

> "When Thane retired from the chair of anatomy in 1919, he was followed by that extremely brilliant figure, the late Sir Grafton Elliot Smith. His arrival at University College sealed my fate. I could not desert anatomy, for I could not leave Elliot Smith. Egyptology, neurology, and anthropology were opened out to me. I saw at close contact the vagaries, turns, twists, and intense driving force of one of the best intellects of the century. Yet I continued to spend almost half of my time in hospital and was mainly interested in the babies and children, for the growth of animal or plant always attracted me. I spent several years in the dual role of anatomist and physician, demonstrator of anatomy and curator of the museum on one side of Gower Street, and assistant to the medical unit on the other."
>
> (Harris, 1951, p.430)

Harris lines are transverse lines of higher density or radio-opaque lines in the metaphyses that can be seen in the radiographs of long bones (usually the tibia, femur and radius) (Roberts and Manchester, 1995, p.175). As the femur and the tibia are fast growing bones in length, they tend to display visible distance between the transverse lines which allows more accurate examination (Hunt and Hatch, 1981, p.466). The femur and tibia are also strong bones that are likely to be the best preserved from an archaeological context (Hunt and Hatch 1981, p.466).

Harris lines can be also detected in other parts of the skeleton: a Polaroid X-ray image of a mummified child showed Harris lines in the illium (Conlogue et al., 2004, p.257). Similar lines seen across the enamel of the teeth are called enamel hypoplasia (Roberts and Manchester, 1995, p.58).

Researchers have studied the formation of Harris lines and dental hypoplasia in the skeletal remains of different populations to establish if they were caused by the same type of stress but the results were inconclusive (Roberts and Manchester, 1995, p.60). For example a study on prehistoric Native Americans proved that Harris lines and dental hypoplasia are not correlated (McHenry and Schulz, 1976, p.509), while another study of the ancient Nubian population found strong correlation between two stress indicators: dental hypoplasia and pitting in the orbital roofs which is known as cribra orbitalia, indicating iron deficiency (Roberts and Manchester, 1995, p.60, 167).

In 2008 a study was carried out by Ritzman et al. to compare all the published methods of estimating the age of formation of the hypoplasia that appear on the enamel. These lines provide evidence of periods of arrested growth and can be compared to Harris lines on the long bones.

Harris lines were found to have no relation to the femoral length, as reported by Mays (1985, p.207). This also highlights the difficulty in establishing a reason for the formation of the lines which were referred to, along with dental hypoplasia, as 'nonspecific indicators' by Buikstra and Cook (1980, p.436) who suggested that, as the lines do not affect the growth of the long bones as demonstrated by Mays (1985, p.207), they should not be referred to as 'arrested growth lines'.

The transverse lines would be visible only in children or sub-adult skeletons as the lines disappear in most adults due to constant remodelling of the bone throughout life (Roberts and Manchester, 1995, p.176). This would explain why the radiographs that show Harris lines usually belong to children or sub-adults, generally between the age of 6 and 12 according to Pointek et al. (2001, p.33). When an adult skeleton does not show evidence of Harris lines, this could indicate either that the lines existed during childhood and then disappeared during adulthood due to bone remodelling, or that the lines did not form during childhood in the first place.

However, Roberts and Manchester (1995, p.176) and Hunt and Hatch (1981, p.466) mention that the lines represent bone lattice or plates formation during a period of time when the bone did not grow in length. This would generally indicate periods of ill health or malnutrition that did not continue for a long time but were followed by a period of health (Roberts and Manchester, 1995, p.176). Poor diet and metabolic stress has been suggested by Hummert and Van Gerven (1985, p.305) as a reason for the formation of the arrested growth lines.

On the other hand, Strouhal and Vyhnánek (1974, p.128) disagree with this idea and suggest that the lines were formed during childhood because of disease and have no relation to poor diet. These lines are common in Egyptian mummies and have been detected in a large number of radiographic investigations (Gray, 1973, p.52; Strouhal and Vyhnánek, 1974, p.128). Other physical anthropologists disagree with the previously mentioned theories regarding the reason behind the formation of Harris lines.

Studies into the formation of Harris lines in animals provided evidence that associates the development of the lines with deficiency in protein and vitamin A (Gronkiewisz et al., 2001, p.46).

Harris lines do not appear on the bones of individuals who suffered long term malnutrition or did not survive a long term illness (Roberts and Manchester, 1995, p.176). The lines were called 'temporary' arrested growth lines by Anderson et al. (1963, p.9); however, Wells (1963, p.406) described them as 'scare lines' and mentioned that they are evidence of stress, and that they remain permanently in the bones. The lines also indicate that recovery took place according to Roberts and Manchester (1995, p.176). For this reason, the lines of arrested growth cannot be considered as associated with death or morbidity. They only indicate that the individual suffered from periods of stress during childhood. However, Brothwell (2008, p.124) would confirm the lines as 'long-term environmental stress'.

In 2007 Metcalfe mentions that the lines *"signify bone regrowth after temporary cessation of longitudinal growth. Merely by measuring the sum total of these lines between the diaphysis and epiphysis of long bones, estimates on the malnutrition and disease status during the childhood of ancient populations can be made"* (Metcalfe, 2007, p.655).

It has been suggested that the lines do not indicate ill health or poor diet but simply the formation of new bone cells while the individual is growing (Antoine, 2010). Roberts and Manchester (1995, p.201) concluded that the reasons behind the stress which may cause formation of the lines are still unknown.

Emotional stress was also mentioned as a possible reason for the formation of Harris lines and dental hypoplasia (Roberts and Manchester, 1995, p.164).

Physical anthropological standards have been developed to evaluate these lines and estimate the age of the individual when each one was formed (Hunt and Hatch, 1981, p.461, Maat, 1984, p.291). Hunt and Hatch (1981, pp.466-467) used a method of estimating the age by measuring the distance between the centre of ossification or the origin of the long bone and each transverse line in the distal and proximal end of the femur and the tibia. They used the same technique to estimate the age at death from the length of long bones. This method is based on growth rates at both ends of each long bone. These rates were measured in 1963 by Anderson et al. who conducted a study of more than 200 sub-adults with Harris lines.

Following a comparative study of two populations, medieval and contemporary in Central Europe, Ameen *et al.* (2005, p.279) related the lines to the high mortality rates and concluded that they were associated with poor diet and unhealthy living standards in Medieval children.

A more recent study on 241 tibiae from Swiss skeletal remains by Papageorgopoulou *et al.* (2011) who formulated a new semi-automated technique to detect and analyse the lines, concluded that diet and ill health are not associated with the formation of the lines which they attributed to normal growth patterns.

Harris lines and ancient Egyptian mummies

In 1897, Flinders Petrie contacted the anatomy department at University College London concerning the possibilities of X-raying some of the mummified remains from the Old Kingdom that he had discovered earlier in Egypt (Taconis, 2005, p.43; Fiori and Nunzi, 1995, p.68). Petrie was studying the possibilities of cannibalism in ancient Egypt and hoped that X-raying these human remains would provide supporting evidence (Taconis, 2005, p.43). These bones were removed from the body in antiquity and wrapped separately in linen bandages (Böni et al., 2004, p.207). The cannibalism assumption was supported by Petrie's observation of teeth marks on human tissue and missing marrow from long bones discovered at Tomb 5 from Naqada and from other sites such as Gerza (Aufderheide, 2003, p.220). The following year, a mummy and more human remains from the cemetery of Deshasheh were X-rayed and the pictures were published in 1898. One of these radiographs shows Harris lines in the distal end of the tibia (Gray, 1967, p.34).

Another example is PUM II which was originally part of the Pennsylvania Museum of Art Collection and in 1934 was sent to the Pennsylvania University Museum as a long term loan (Angel and Zimmerman, 1982, p.122). Cockburn corresponded with O'Connor, the curator of the Pennsylvania University Museum who agreed to send the mummy to Detroit for autopsy (Cockburn et al., 1975, p.1160). The autopsy took place at the Department of Physiology, Wayne State University School of Medicine, Detroit, Michigan (Lawrence, 1980, p.362). A hammer and a chisel were used to break through the hard resinous layers during the 8 hour mummy unwrapping (Cockburn, 1973, p.470). Radiographic examination was carried out before the autopsy to assist in the process of choosing the perfect candidate for the unwrapping as two other mummies were also available (Cockburn et al., 1975, p.1155). X-ray images showed that the deceased suffered from osteomyelitis in the right fibula which caused a deformation of the leg (Cockburn, 1973, p.470). After the autopsy of the mummy, radiographic investigation was also used to examine several pathological features and skeletal abnormalities such as an extra lumbar vertebra and the detection of Harris Lines in the long bones (Cockburn et al., 1975, pp.1155-1157).

In 1974 Cockburn, who worked for the Department of Anthropology of the Smithsonian Institution at that time, carried out another autopsy on a mummy, dating to the 20th Dynasty, from the collection of the Royal Ontario Museum, Toronto (Hart et al., 1977, p.461). The mummy, known as Nakht, a 16 year old boy, was X-rayed before the unwrapping (Hart et al., 1977, p.461). Xeroradiographic and tomographic images were also obtained before the autopsy (Rideout, 1977, p.464). The radiological examination suggested that Nakht suffered from serious illness during the last two years of his life, as Harris lines were detected in the images which also showed that remains of the liver and lung tissue were still in the body (Rideout, 1977, p.464). The age at death and the sex of the mummy were assessed from the X-rays, which also showed the absence of evidence for mummification; the body was preserved by natural dehydration (Rideout, 1977, p.464). The results of the studies carried out by Cockburn and his team were published in 1983 in *Mummies, Disease and Ancient Cultures* (Cockburn et al 1983).

Recent research

During my research on a selected group of Third Intermediate Period mummies at the British Museum, plain radiographic digital images of the lower extremities were used to identify Harris lines in the specimens. Harris lines were reported from two mummies (Tjayasetimu EA 20744, Tjentmutengebtiu EA 29577).

EA 20744: Mummy and cartonnage case of Tjayasetimu

This mummy was acquired by the British Museum in 1888. Taylor (2010, p.89) suggests that the mummy belongs to the 22nd Dynasty. It was on display in the First Egyptian Room (Case K) in 1898 when the guide to the First and Second Egyptian Rooms was compiled by Budge (1898, p.28). He mentioned that the cartonnage was closed at the back using the usual lacing technique and a wooden board was attached to the foot area to secure the mummy inside the case. Budge (1898, p.28) also mentioned that the mummy was found at Deir El Bahari.

Dawson and Gray (1968, p.19) reported that the mummy belongs to a 12 year-old girl, based on the examination of the unerupted teeth. They also reported that the mummy does not display any fractures or dislocations. The X-rays showed that the thoracic cavity did not contain any packages and appears to be empty. Apart from some Harris lines on both ends of the tibia, Dawson and Gray (1968, p.20) did not mention any abnormalities detected during the radiological examination.

During the recent examination of the mummy it has been reported that the long bones are in a good state and intact. There is evidence of open growth plates confirming the skeleton of a child. There is no evidence of joint or bone disease. There are prominent 'Harris growth arrest lines' in the distal tibiae which might indicate some cause of interruption of enchondral ossification at that time of skeletal development or it could be part of a normal growth pattern.

EA 22939: Mummy and cartonnage case of Tjentmutengebtiu

The mummy was purchased by the British Museum from Raymond Sabatier in 1890, according to the British Museum online database. It dates back to the early 22nd Dynasty (Taylor, 2009, p.402) and was found at Thebes. Tjentmutengebtiu was a singer and priestess in the Temple of Amun ($nb(t)$ pr $šm^cyt$ n imn) while her father Khonsmes was a *wab* priest (w^cb n imn) in the temple of Amun (Dawson and Gray, 1968, p.8, Budge, 1898, p.28). Her mother's name was also mentioned on the case as Mehenmut(em)hat (Dawson and Gray, 1968, p.8). The mummy was on display in case K in the First Egyptian Room in 1898, together with Tjayasetimu (EA 20744) (Budge, 1898, p.28).

Artificial eyes still in place in the orbits and a number of amulets were visible in the radiographs. They also reported an oval plate incised with a *bnw* bird, a heart scarab and an incision plate which they suggested could have been made of metal or gilded wood (Dawson and Gray, 1968, p.8). Subcutaneous packages and a number of fractures, although they did not suggest if they were post-or ante-mortem were observed in the X-ray images. They also noted a number of Harris lines on the tibia which they suggest indicates periods of ill health during childhood (Dawson and Gray, 1968, p.8). However, during my research no evidence of 'Harris growth arrest lines' were now detected from the X-ray images.

EA 29577: Cartonnage case and mummy of Djedameniufankh

The coffin and the cartonnage case were purchased by Budge in 1897 and were displayed in two separate cases in the museum in 1898 (Budge, 1898, p.45, 68). Budge (1898, p.45, 68) mentioned that the coffin and mummy belong to the 26th Dynasty and that they were found in Thebes, while Dawson and Gray (1968, p.12) suggest that the ensemble date back to the 21st Dynasty and that it was disturbed, looted, restored and rewrapped before it was placed inside the cartonnage. However, Taylor (2010, p.90) dates the mummy to the 22nd Dynasty.

The mummy which was thought to be of a female according to Budge (1898, p.45, 68) was found to be of a young male adult by Dawson and Gray (1968, p.12) who discovered the poor state of preservation, disturbance and disarticulation of the skull and upper body. It appeared from the X-ray images that the head was upside down with a tooth and one artificial eye inside the skull. They reported that the mandible was missing. Lines of arrested growth were detected in both femora. The radiographs showed that the lower part of the body was in a better condition with articulated long bones (Dawson and Gray, 1968, p.12).

During the recent examination, the skeleton was found to be that of an adult in a poor state of preservation, with extensive postmortem disruption. There is a suggestion of osteopenia with prominent vertical trabeculae in the vertebral bodies. There are Harris lines visible in the tibiae. There are two supporting struts (papyrus stems?) evident which extend from torso to feet: the one on the right is longer than that on the left. Widening of the diploic space of the skull vault raises the possibility of haemolytic anaemia or anaemia with marrow hyperplasia.

To date, there has been too little in depth study of the Harris lines detected on the skeletal remains from ancient Egypt which highlights the potential for further study and analysis.

For more information about Harris lines in archaeological context, see Pointek et al., 2001, pp.33-43, for evidence of Harris lines correlating with seasonal periods of poor diet, see Lobdell, 1984, pp.109-116, and on formation and persistence of the lines, see Hummert and Van Gerven (1985, pp.297-306). For methods of estimating the age of formation of dental hypoplasia lines see Ritzman et al., 2008, pp.348-361.

Acknowledgments

I would like to thank Dr. John Taylor, The British Museum, Prof. Rosalie David, The University of Manchester and Prof. Judith Adams for their vital support during this research. I am also grateful to the KNH Centre and the Faculty of Life Sciences for granting me the Faculty Bursary Award which allowed me to continue my research.

References

Ameen, S., Staub, L., Ulrich, S., Vock, P., Ballmer, F. and Anderson, S. E., 2005. Harris lines of the tibia across centuries: a comparison of two populations, medieval and contemporary in Central Europe. *Skeletal Radiology*, 34 (5), pp.279-284.
Anderson, M., Green, W.T. and Blais Messner, M., 1963. Growth and Prediction of Growth in the Lower Extremities. *Journal of Bone Joint Surgery*, 45-A, pp.1-14.
Angel, J.L. and Zimmerman, M.R., 1982. T. Aidan Cockburn, 1912-1981: A Memorial. *American Journal of Physical Anthropology*, 58, pp.121-122.
Antoine, D., 2010. *Personal Communication with the Assistant Keeper*. Physical Anthropology, Department of Ancient Egypt and Sudan, the British Museum.
Aufderheide, A.C., 2003. *The Scientific Study of Mummies*. Cambridge: Cambridge University Press.
Böni, T., Frank J., Rühli, F.J. and Chhem, R.K., 2004. History of Paleoradiology: Early Published Literature, 1896–1921. *Journal of the Canadian Association of Radiologists*, 55, pp.203-210.

Brothwell, D.R., 2008, Paleoradiology in the Service of Zoopaleopathology. In R. K. Chhem and D. R. Brothwell, eds. *Paleoradiology: Imaging mummies and fossils*. Berlin: Springer-Verlag. pp.119-145.

Budge, E.A.W., 1898. A *Guide to the First and Second Egyptian Rooms: Mummies, Mummy-Cases and Other objects Connected with the Funeral Rites of the Ancient Egyptians*. Reprint 2009. London: Read Books Ltd.

Buikstra, J.E. and Cook, D.C., 1980. Palaeopathology: An American Account. *Annual Review of Anthropology*, 9, pp.433-470.

Cockburn, A., 1973. Death and Disease in Ancient Egypt. *Science*, 181(4098), pp. 470-471.

Cockburn, A., Barraco, R.A., Reyman, T.A. and Peck, W.H., 1975. Autopsy of an Egyptian Mummy. *Science,* 187(4182), pp.1155-1160.

Cockburn, A., Cockburn, E., Reyman, T. A., (eds) 1983. *Mummies, Disease and Ancient Cultures*. New York: Cambridge University Press.

Conlogue, G., Nelson, A. and Guillén, S., 2004. The Application of Radiography to Field Studies in Physical Anthropology. *Journal of the Canadian Association of Radiologists*, 55, pp.254-257.

Dawson, W.R. and Gray, P.H.K., 1968. *Catalogue of Egyptian Antiquities in The British Museum I: Mummies and Human Remains*. London: The Trustees of the British Museum.

Fiori, M.G. and Nunzi, M.G., 1995. The Earliest Documented Applications of X-rays to Examination of Mummified Remains and Archaeological Materials. *Journal of the Royal Society of Medicine*, 88, pp.67-69.

Gray, P.H.K., 1967. Radiography of Ancient Egyptian Mummies. *Medical Radiography and Photography*, 43, pp.34-44.

Gray, P.H.K., 1973. The Radiography of Mummies of Ancient Egyptians. *Journal of Human Evolution*, 2, pp.51-53.

Gronkiewisz, S., Kornafel, D., Kwiatkowska, B. and Nowakowski, D., 2001. Harris's Lines Versus Children's Living Conditions in Medieval Wrocław, Poland. *Variability and Evolution*, 9, pp.45-50.

Harris, H.A., 1926. The growth of the long bones in childhood: with special reference to certain bony striations of the metaphysis and to the role of the vitamins. *Archives of Internal Medicine*, 38(6), pp.785-806.

Harris, H.A., 1931a. Lines of Arrested Growth in the Long Bones in Childhood: the Correlation of Histological and Radiographic Appearances in Clinical and Experimental Conditions. *British Journal of Radiology*, 4, pp.622-640.

Harris, H.A., 1931b. Lines of arrested growth in the long bones of diabetic children. *British Medical Journal*, 3668, p.700.

Harris, H. A., 1951. My Academic Life. *British Medical Journal,* 4729, pp.429-431.

Hart, G.D., Cockburn, A., Millet, N.B. and Scott, J.W., 1977. Lessons Learned from the Autopsy of an Egyptian mummy. *Canadian Medical Association Journal* 117, p.461.

Hummert, J.R. and Van Gerven, D.P., 1985. Observations on the Formation and Persistence of Radiopaque Transverse Lines. *American Journal of Physical Anthropology* 66, pp.297-306.

Hunt, E.E. and Hatch, J.W., 1981. The Estimation of Age at Death and Ages of Formation of Transverse Lines from Measurements of Human Long Bones. *American Journal of Physical Anthropology* 54, pp.461-469.

Lawrence, S.V., 1980. Unravelling the Mysteries of the Mummies. *Science News (Washington, D. C.)*, 118, pp.362-364.

Lobdell, J.E, 1984. Harris Lines: Markers of Nutrition and Disease at Prehistoric Utqiagvik Village, The Frozen Family from the Utqiagvik Site, Barrow, Alaska: Papers from a Symposium. *Arctic Anthropology*, 21, pp.109-116.

Maat, G.J.R., 1984. Dating and Rating of Harris's Lines. *American Journal of Physical Anthropology* 63, pp.291-299.

Mays, S.A., 1985. The Relationship Between Harris Line Formation and Bone Growth and Development. *Journal of Archaeological Science*, 12, pp.207-220.

McHenry, H.M., Schulz, P.D., 1976. The association between Harris lines and enamel hypoplasia in prehistoric California Indians. *American Journal of Physical Anthropology*, 44, pp.507-512.

Metcalfe, N.H., 2007. A description of the methods used to obtain information on ancient disease and medicine and of how the evidence has survived. *Postgraduate Medical Journal*, 83, pp.655-658.

Papageorgopoulou, C., Suter, S.K., Rühli, F.J., Siegmund, F., 2011. Harris lines revisited: prevalence, comorbidities, and possible etiologies. *American Journal of Human Biology*, 23(3), pp.381-91.

Pointek, J., Jerszyñska, J. and Nowak, O., 2001. Harris Lines in Sub-adult and Adult Skeletons from the Mediaeval Cemetery in Cedynia, Poland. *Variability and Evolution*, 9, pp.33-43.

Rideout, D.F., 1977. Radiologic Examination. In G.D Hart, A. Cockburn, N.B. Millet and J.W Scott. Autopsy of an Egyptian Mummy. *Canadian Medical Association Journal*, 117, pp.461-476.

Ritzman, T.B., Baker, B.J. and Schwartz, G.T., 2008. A Fine Line: A Comparison of Methods for Estimating Ages of Linear Enamel Hypoplasia Formation. *American Journal of Physical Anthropology*, 135, pp.348–361.

Roberts, C. and Manchester, K., 1995. *The Archaeology of Disease*. Ithaca, New York: Alan Sutton Publishing Ltd.

Seipel, W., 1996. Research on Mummies in Egyptology. An overview. In K. Spindler, H. Wilfing, E. Rastbichler-Zissernig, D. zur Nedden and H. Northdurfter, eds. *Human Mummies, A Global Survey of their Status and the Techniques of Conservation*. Vienna: Springer-Verlag. pp.41-45.

Smith, G.E., 1926. The significance of Anatomy. *The Lancet*, 5384, pp.943-946.

Strouhal, E. and Vyhnánek, L, 1974. Radiographic Examination of the Mummy of Qenamun the Seal-Bearer. *Zeitschrift fur agyptische Sprache und Altertumskunde*, 100, pp.125-129.

Taconis, W.K., 2005. Mummification in Ancient Egypt with a History of the Investigation of Egyptian Mummies. In M.J. Raven and W.K. Taconis, eds. *Egyptian Mummies, Radiological Atlas of the Collections in the National Museum of Antiquities in Leiden*. Turnhout, Belgium: Brepols Publishers. pp.35-51.

Taylor, J. H., 2010. *Egyptian Mummies*. London: The British Museum Press.

Taylor, J.H., 2009. Coffins as Evidence for a 'North-South divide' in the 22nd-25th Dynasties. In G.P.F. Broekman, R.J. Demarée and O.E. Kaper, eds. *The Libyan Period in Egypt, Historical and Cultural Studies into the 21st– 24th Dynasties: Proceedings of a Conference at Leiden University, 25-27 October 2007.* Leiden: Peeters Leuven. pp.375-415.

Wells, C., 1963 The Radiological Examination of Human Remains. In D. Brothwell and E. Higgs, eds. *Science in Archaeology, A Comprehensive Survey of Progress and Research*. London: Thames and Hudson. pp.401-412.

An interesting example of a condylar fracture from ancient Nubia suggesting the possibility of early surgical intervention

Mervyn Harris,[1] Tristan Lowe[2] and Farah Ahmed[3]

[1]KNH Center for Biomedical Egyptology, The University of Manchester, Manchester, UK
[2]Henry-Mosley Imaging Facility, School of Materials, The University of Manchester, Manchester, UK
[3]Micro-CT Specialist, The Natural History Museum, London, UK

Abstract

The specimen of a skull from a cemetery in Qurta, ancient Nubia from the Duckworth Laboratory at the Leverhulme Centre for Human Evolutionary Studies (LCHES) in Cambridge demonstrated an old healed fracture of the mandibular condyle on the right side. There was no evidence of infection, but the normal anatomy of the condylar head had been destroyed and part of the condylar head was also ankylosed to the glenoid fossa. Associated with the injury was what appeared to be a man-made linear incision and a cruciate bony deficiency which also appeared to be man-made. Micro-CT and 3D surface laser scanning of the linear 'incision' confirmed the peri-mortem nature of the lesion. The fracture, morphology and anatomical position of the man-made lesions suggest that some form of surgical intervention may have been employed in order to re-establish a degree of jaw movement.

Introduction

Specimen NU761 from cemetery 116, grave 5 (Qurta) housed in the Duckworth Laboratory at the LCHES in the University of Cambridge demonstrates an example of a mandibular condylar fracture. The injury was old, as evidenced by the degree of remodelling and new bone formation that had taken place. No evidence of infection was visible and part of the condylar head was ankylosed to the glenoid fossa. Hypertrophic bone growth both at the angle of the mandible on the right side and in the anterior region of the body of the mandible corresponding to the area of the chin was noted. Two areas of interest in the region of the glenoid fossa appeared to have been man-made, suggesting that of some form of surgical intervention may have taken place.

Materials

The specimen consisted of a complete adult skull. The pronounced areas of attachment of the major muscles of mastication (masseter and temporalis), together with the heavy mandibular morphology suggested that the individual was male.

Methods

The skull was digitally photographed and micro-CT analysis was carried out at The Natural History Museum, London (NHM) using a Nikon X Tec 225K micro-CT scanner. The images were viewed and 3D reformation was carried out using OsiriX Imaging Software and a DICOM Viewer. The surface of the lesion, suggestive of surgical intervention, was scanned at the NHM using an Alicona surface laser scanner (Alicona GmbH Schonau Germany).

Results

Examination of the micro-CT scans revealed an old injury to the right mandibular condyle. There was no evidence of infection but the normal anatomy of the condylar head was destroyed. Part of the condylar head was fractured, separated from the rest of the condyle and ankylosed to the glenoid fossa. Hypertrophy of the bone at the angle of the mandible on the right side was noted, together with a degree of hypertrophic bone to the right of the anterior portion of the body of the mandible in the region of the chin. Proximal to the right external auditory meatus were two lesions. One had the appearance of a cruciate deficiency in the bone. Above this and parallel to it, was a second horizontal lesion suggesting that some form of bladed surgical instrument may have been used.

Surface laser scanning of the lesion using an Alicona laser scanner confirmed what was visible via macroscopic examination and photography, namely that the lesion was relatively smooth sided, wider at the surface and tapering to the bottom, suggesting that it had been made by a double sided bladed instrument such as a small sharp knife. Micro-CT and 3D digital reformation confirmed what was seen on laser scanning, but in addition added visible evidence of cut marks on several areas on both sides of the walls of the lesion.

Discussion

Mandibular fractures usually occur as a result of direct force trauma, normally at the point of impact. Indirect fractures can also occur in other parts of the mandible through the transmission of force to a weak point, e.g., the condylar neck. One example might be a punch to the chin resulting in a fracture of either a single or both condyles (De Luca et al., 2011). Numerous sequelae can result from this type of injury. In the immediate post traumatic phase, muscle spasm combined with pain and swelling would inevitably result in restricted mandibular movement. In the absence of treatment, the inevitable sequelae to this type of injury are mandibular deviation, masticatory difficulties and varying degrees of malocclusion (Valiate et al., 2008)

Anatomically, the mandible varies greatly in thickness as does its strength and ability to resist trauma. The body of the mandible is the thickest and most resilient area and the ramus and condylar necks are the weakest. Condylar fractures can be situated outside the fibrous capsule surrounding the joint and are referred to as extra-capsular (condylar neck or sub condylar) or situated higher up within the fibrous capsule (intra-capsular). In addition, fractures can be non-displaced or displaced, deviated or dislocated (Zachariades et al., 2006)

Damage to the condylar area has a greater effect on movement than a fracture to the body of the mandible as the temporomandibular joint (TMJ) itself is directly affected. In the absence of immobilisation of the fracture and restoration of something approaching a normal anatomy, mandibular function is affected to a greater or lesser extent depending on the site and extent of the injury. Deviation of the mandible and malocclusion will result (De Souza et al., 2007; Valiati et al., 2008), the long term consequences of which would be pathological changes in the TMJ such as osteoarthritis and possibly osteonecrosis if the blood supply is compromised. Ankylosis of the TMJ on the affected side can also occur (Dongmei, 2008), as well as degenerative osteoarthritis of the TMJ on the opposite side caused by excessive load being placed on the non-fractured joint.

Surgical treatment of a mandibular condylar fracture is a relatively complex procedure, even by present day standards. Nevertheless, the specimen in question appears to suggest that some form of surgical intervention may have been attempted - possibly to separate an ankylosed area of bone in order to allow some degree of movement of the jaw.

The Edwin-Smith Papyrus, the most detailed surgical papyrus to survive from ancient Egypt deals with mandibular injuries and distinguishes between fracture and dislocation. Fracture especially in the presence of fever and/or infection is classified as an injury not to be treated, whereas there are clear instructions for the treatment and reduction of dislocations of the mandible, although this is not the same as surgical intervention.

Damage sustained to the TMJ resulting in ankylosis of the condylar head to the glenoid fossa (figures 1 and 2) would have made mastication difficult. In the initial period after the injury, there would have been no contact with the glenoid fossa between the ramus and condyle on the affected side. In order to attempt to re-establish contact and allow

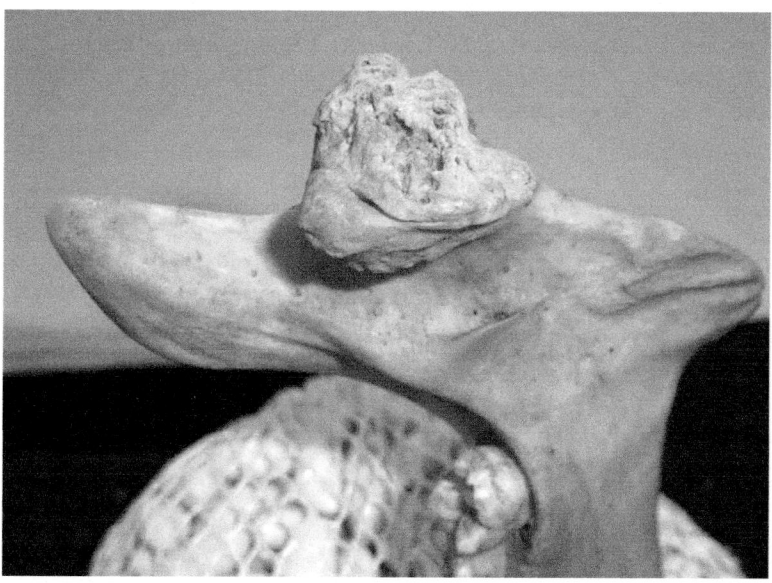

Figure 1: Condylar head showing extensive destruction of the normal anatomy of the condylar head. (Courtesy of the Duckworth Laboratory, the University of Cambridge)

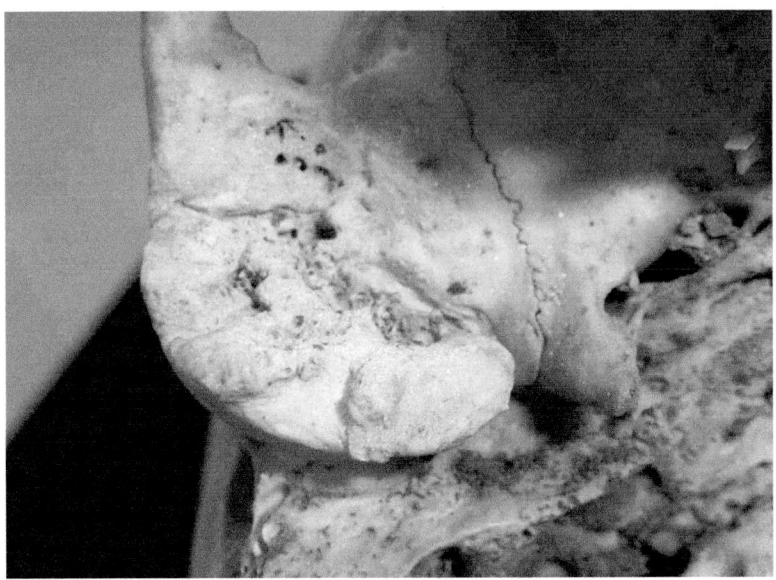

Figure 2: View looking directly into the Glenoid fossa showing bone from the condylar head ankylosed to the fossa. (Courtesy of the Duckworth Laboratory, the University of Cambridge)

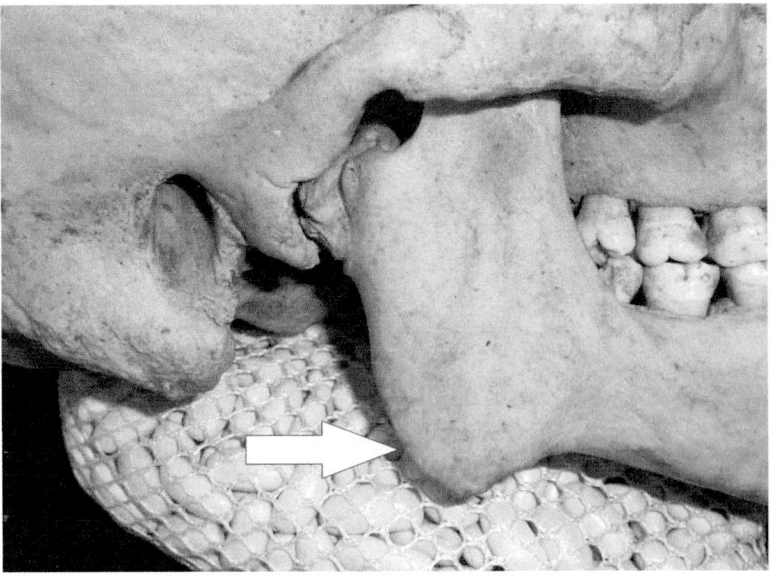

Figure 3 Hypertrophic bone of mandibular ramus.
(Courtesy of the Duckworth Laboratory, the University of Cambridge)

some degree of masticatory function, the bone at the angle of the mandible appears to have increased in length thereby pushing the mandibular ramus up to a degree sufficient to re-establish contact (figure 3).

Close examination of the skull revealed two areas of interest associated with the fracture. What appeared to be a straight incision in the bone made by a sharp bladed instrument was visible anterior to the external auditory meatus at the beginning of the zygomatic arch. The sides of the 'incision' appeared to be corticated, indicating that it was peri-mortem in nature and tapered to a point at the bottom as one would expect from an incision made by a sharp bladed instrument (figure 4, white arrow). The perfectly straight nature of the lesion made it most unlikely that this could have been the result of some natural injury or abnormality. Above this was a second lesion with a cruciate vertical and horizontal component, both of which were relatively, but not completely, straight, suggestive of the type of deficiency in the bone which one might expect to see in a present day case if a flat surgical elevator had been used to elevate and separate two areas of bone and fibrous tissue (figure 4, black arrow). Although there is no evidence of such an instrument being available in antiquity, the shape of the deficiency in the bone suggests that something akin to a type of flat instrument or blade had been inserted.

Surface scanning of the straight 'incision' mark with the Alicona laser scanner confirmed the tapering and smooth sided nature of the lesion, in keeping with something produced by a sharp double sided bladed instrument (figure 5). Micro CT was also carried out,

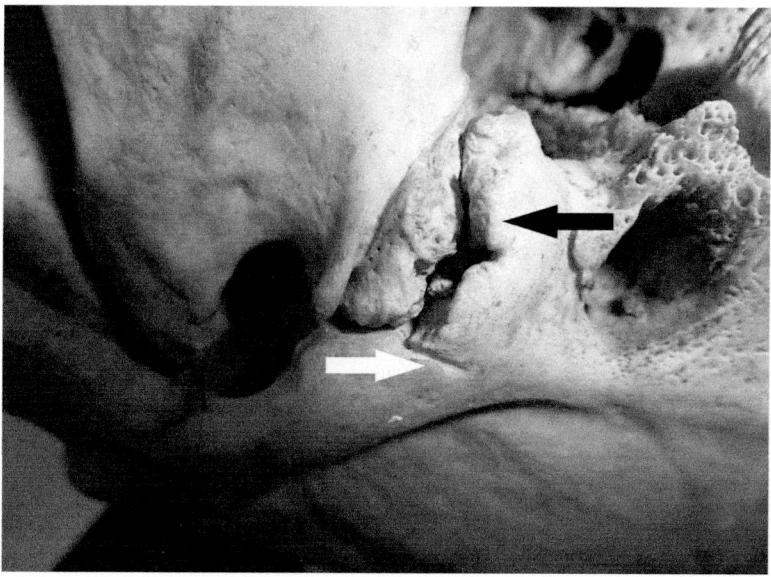

Figure 4: Demonstrating a sharp incision
— white arrow and a second horizontal entry mark – black arrow).
(Courtesy of the Duckworth Laboratory, the University of Cambridge)

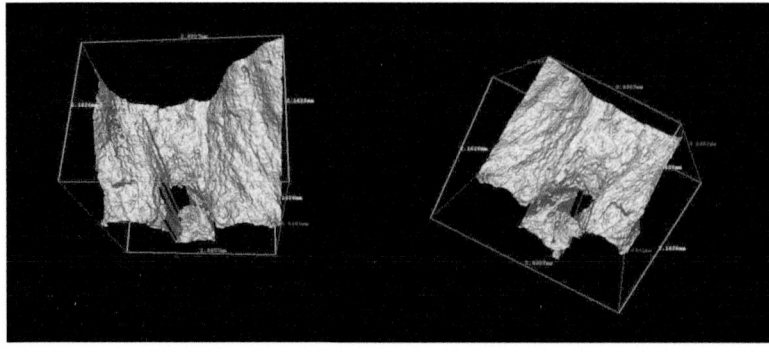

5A 5B

Figure 5 A and B: Alicona surface laser scan
of the area indicated by the white arrow in figure 6 showing
smooth and tapered sides of the 'incision'.

followed by 3D digital reformation at the Henry Mosley X-ray imaging facility at The University of Manchester.

This confirmed the smooth sided and tapered nature of the lesion, but in addition, several horizontal cut marks on the side walls (figure 6) demonstrating that the cut had not been made with one thrust of a blade but rather, had been made employing a careful and deliberate cutting action. A section through the lesion also revealed it to have penetrated the cortex, extending into the cancellous bone below.

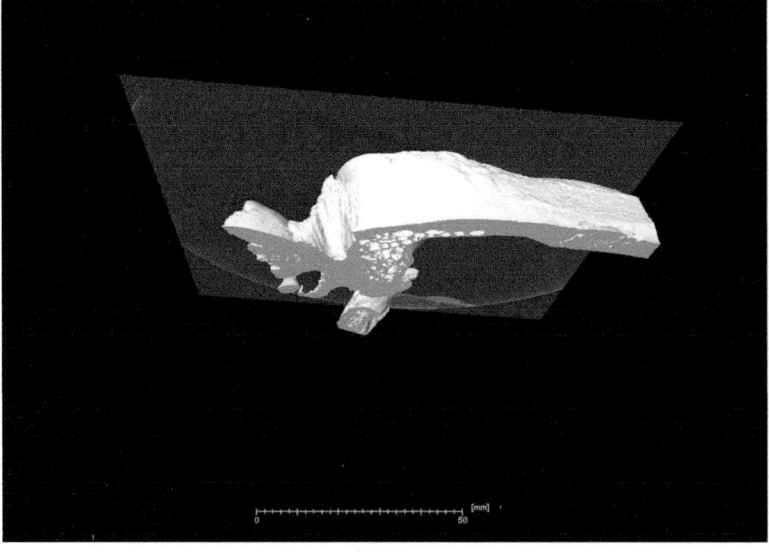

Figure 6: 3D reformation of Micro CT
showing horizontal cut marks at various levels along the side of the lesion.

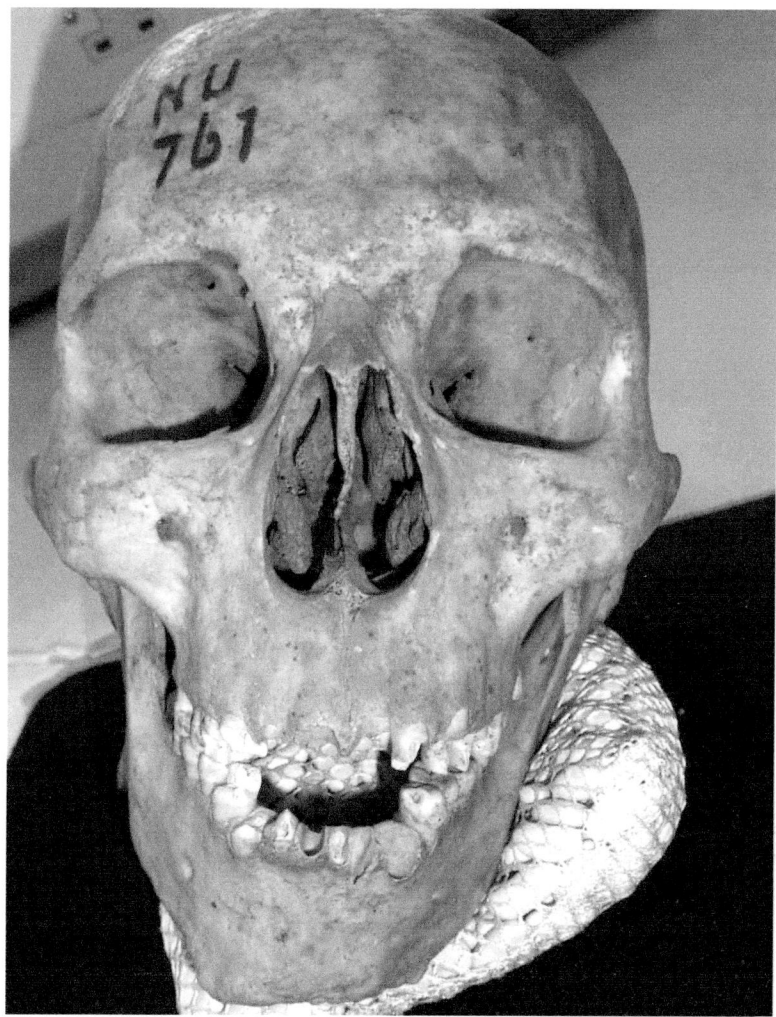

Figure 7: Anterior view of skull with mandible articulated.
A degree of facial asymmetry towards the right can be seen and enlargement of the bone at the anterior portion of the body of the mandible, in the region of the chin.
(Courtesy of the Duckworth Laboratory, the University of Cambridge)

Having established that the lesion in question is most likely to be a man-made incision, for what reason would such an incision have been carried out? At a basic level, TMJ fractures can be broadly classified as intra or extra-capsular. Intra-capsular fractures lying within the fibrous capsule surrounding the joint when untreated often result in ankylosis of the joint and there is evidence that that has happened in this case. Severe restriction in jaw movement, masticatory difficulty, uneven tooth wear and possible facial asymmetry are likely consequences and evidence of a degree of facial asymmetry can be seen in figure 7.

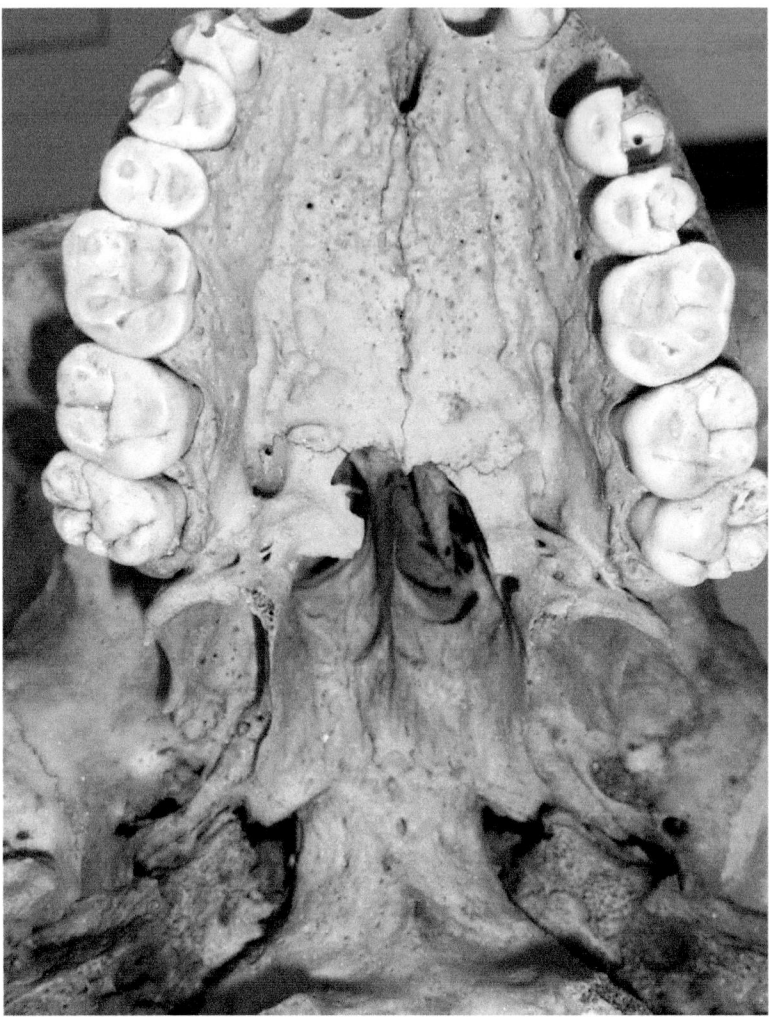

Figure 8: Occlusal surface of molar teeth demonstrating even occlusal wear
on both sides of the dental arch.
(Courtesy of the Duckworth Laboratory, the University of Cambridge)

Uneven tooth wear resulting from long term masticatory difficulty would also be a likely consequence, but evidence of this is absent in this case (figure 8). There is certainly evidence of occlusal attrition on the premolar and molar teeth on both sides, but the attrition is evenly distributed and in line with what one normally expects to see on such an ancient skull. The degree of occlusal attrition present also gives some indication of the age at death of the individual. Because the upper first molar erupts first, the degree of attrition is greater than that of the second molar which erupts later. Finally, the third molar is not fully erupted until the age of 20-21 years and one can see that there is very little evidence of occlusal attrition associated with the third molars. This suggests an age at death of around mid-late twenties.

There are two possible explanations for the uniform occlusal wear pattern observed. The hypertrophy of the bone at the angle of the mandible on the side of the fracture (figures 5A and B) would have increased the length of the ramus on that side, thereby restoring the occlusion to a degree sufficient to allow even articulation of the premolar and molar teeth although mandibular movement would still have been severely restricted. The second explanation is that movement and articulation was restored by surgically dividing the fibrous tissue surrounding the joint and separating the ankylosed bone by elevating them apart, which would account for the cruciate deficiency seen (figure 6). This combination of increased length of the mandibular ramus on the affected side and surgical intervention to restore free movement could be sufficient to restore something akin to a normal occlusion.

Also absent is evidence of the degenerative arthritis of the temporomandibular joint (TMJ) of the opposite side which would inevitably accompany long term occlusal imbalance. Once again, for whatever reason, occlusal balance had been restored to a degree sufficient to prevent such an occurrence.

The final question is whether or not TMJ surgery is a reasonable hypothesis for the period. Surgery in this anatomical position is potentially hazardous due to the presence of a number of vital structures. Anterior to the TMJ lies the parotid salivary gland and within this, on the posterior aspect, lies the superficial temporal artery, external carotid artery and the facial nerve. Surgery in such a confined area containing a number of vital structures in close proximity to one another is very complicated. Although one might reasonably argue that surgery of this nature would be unlikely in antiquity, the evidence of a man-made incision in exactly the correct anatomical position to avoid all of these vital anatomical structures in order to carry out what would be the appropriate procedure to separate the joint and restore mandibular movement would require a considerable number of coincidences to come together in order to produce such a result.

Conclusion

As is common when dealing with ancient specimens, a definitive diagnosis can be difficult. The specimen clearly demonstrates a long standing intra-capsular fracture of the mandibular condyle. What, on visual examination appeared to be a deliberate incision was supported by micro-CT. The absence of an uneven occlusal wear pattern associated with the TMJ of the opposite side also suggests that effective treatment may have been provided.

The lesions are truly intriguing and suggest that deliberate surgical intervention had taken place in order to treat the individual concerned and it is difficult to suggest any other reason for the bony lesions observed. Although a thrust with a sharp bladed weapon might possibly explain the incision, the successive parallel cuts situated on both sides of the incision as demonstrated on 3D reformation of the micro-CT of the incision line would suggest otherwise. An injury caused by a thrust with a bladed weapon also fails to explain the cruciate lesion associated with the incision.

Acknowledgements

Dr. Marta Lahr, Leverhulme Centre for Human Evolution, University of Cambridge.

This work was supported by The Wellcome Trust [WT090575MA].

References

Breasted, J.H., 1930, *The Edwin Smith Surgical Papyrus*. Chicago: University of Chicago.

De Luca, S., Viciano, J., Irinita, J., Lopez-Lorenzo, S., Botella, D., 2011. Mandibular fracture and dislocation in a case study from the Jewish cemetery of Lucena (Cordoba). *International Journal of Osteoarchaeology* 23 (4), pp. 485-504.

De Souza, M., Oeltjen, J.C., Panthaki, Z.J. and Thaller, S.J., 2007. Posttraumatic Mandibular Deformities. *Journal of Craniofacial Surgery*, 18 (4), pp.912-16.

Dongmei, H.E., Ellis, E., and Zhang, Y., 2008. Aetiology of TMJ Ankylosis Secondary to Condylar Fractures: The role of Concomitant Mandibular Fractures. *Journal of Oral and Maxillofacial Surgery*, 66 (1), pp.77-84.

Romanes, G.J. 1967. *Cunningham's Manual of Practical Anatomy*, 13th Edn. Oxford: Oxford University Press.

Valiati, R., Ibrahim, D., Abreu, M.E.R., Heitz, C., Oliveira, R.B., Pagnoncelli, R.M. and Silva, D.N., 2008. The treatment of condylar fractures; to open or not to open? A critical view of this controversy. *International Journal of Medical Sciences*, 5 (6), pp.313-18.

Zachariades, N., Metizis, M., Mourouzis, C., Papadakis, D., Spanau, A., 2006. Fractures of the mandibular condyle; a review of 466 cases. *Journal of Cranio-Maxillofacial Surgery*, 34, pp.421-32.

An overview of the evidence for tuberculosis from ancient Egypt

Lisa Sabbahy

Department of Sociology, Anthropology, Psychology and Egyptology,
American University in Cairo, Cairo, Egypt

Abstract

The diagnosis of tuberculosis in ancient Egyptian human remains has been largely based on spinal degeneration, or Pott's disease. Since the late 1990's, developments in biomolecular techniques have made possible the retrieval of ancient Mycobacterium complex DNA from ancient bones and tissue, as well as the ability to identify the specific strain of Mycobacterium causing the disease. This paper presents an overview of the evidence for tuberculosis in ancient Egyptian skeletons and mummies from the early 1900's to the present.

Introduction

This paper presents an overview of the study of tuberculosis in ancient Egypt, and how the ability to recognise and diagnose tuberculosis from ancient human remains has changed over time. In particular, since DNA was first retrieved from an ancient Egyptian mummy in 1985 (Pääbo, 1985), tremendous progress has been made in not only recognising *Mycobacterium tuberculosis* (TB) DNA in ancient Egyptian human remains, but also particular strains of *Mycobacterium*, and how they may be related, and may have evolved during the time span of ancient Egyptian civilization.

Traditionally, of course, bones are what have been studied for evidence of past disease, and in the case of *Mycobacterium tuberculosis* that means identifying bone lesions. Approximately 4% of the people infected by M*ycobacterium tuberculosis*, in the modern world, develop bone lesions as the bacteria spreads from the lungs in the blood and lymph system into the bone marrow (Brown and Brown, 2011, p. 711). This means that 'only a few cases with bone tuberculosis may indicate a much more widespread epidemiological occurrence of the disease' (Zink et al., 2001, p.355). A diagnosis of tuberculosis is almost always based on lesions in the thoracic and lumbar vertebrae (Brown and Brown, 2011, pp.849-850). In their most destructive form these lesions cause vertebral collapse and fusion, producing the curved spine known as Pott's disease, named after Sir Percival Pott who was a surgeon at St. Bartholomew's Hospital in London, and first described it in 1779.

Physical diagnosis

The first published description of Pott's disease in ancient Egyptian and Nubian remains was by Derry in his 1909 anatomical report in the Archaeological Survey of

Nubia. In the first case in his report he describes the spinal disease in the skeleton of a young woman from a C-Group cemetery near Bab al-Kalabsha. A short note is also added in by G.E. Smith describing the extremely bent spine and psoas abscess of the mummy of the 21st Dynasty High Priest of Amun Nesperehan from Thebes, which Smith and Ruffer published the following year (Derry, 1909, pp.31-32). Nesperehan's case of Pott's disease is probably the most famous from ancient Egypt (Ruffer, 1921, pp.3-10).

Morse, Brothwell, and Ucko summarised the evidence for tuberculosis in ancient Egypt and Nubia in 1964, presenting the evidence from literature, art and skeletal remains. This is still probably the most quoted article on the subject. They state that the cases based on skeletal remains were all identified on 'the basis of involvement of the spine; bone tuberculosis in other locations would be indistinguishable from too many other diseases' (Morse, Brothwell, and Ucko, 1964, p. 528). In all, they give the evidence for 31 cases. In 1988, Strouhal reported on two newly found burials with vertebral tuberculosis, and briefly reviewed all of the evidence for it from Egypt and Nubia again (Strouhal, 1991). He also produced a further update in 1999, including the earliest results of work with *Mycobacterium tuberculosis* DNA (Strouhal, 1999), which will be discussed below.

The evidence used for a tuberculosis diagnosis expanded first in 1979, when Zimmerman published microscopic confirmation of pulmonary TB in the partially preserved post New Kingdom mummy of a child from an intrusive burial at on the West Bank of Thebes. He found 'tubercle bacilli in the vertebral bone' and red blood cells in the trachea and lungs that was 'consistent with fresh and probably fatal hemorrhage' (Zimmerman, 1979, pp.606-607).

Genetic diagnosis

The next development came in the 1990s when work with ancient DNA began to recover DNA from viruses, bacteria and parasites. In 1997 Nerlich and Zink announced the retrieval of a DNA sequence from the lung tissue of a New Kingdom Egyptian mummy of a 35-year old man from the West Bank of Thebes that showed 'homology to the DNA of *M. tuberculosis*'. This molecular evidence backed up a macroscopic examination, which had shown evidence of pulmonary tuberculosis with osseous spread (Nerlich, et al. 1997). A more detailed publication of this study came out in 1999 (Zink et al., 1999)

In 1998 a team, working with Crubezy at Adaima in Upper Egypt, was able to retrieve a DNA fragment from samples of the rib and vertebra of a Predynastic Period child with Pott's disease that was 'sequenced and is consistent with an original *Mycobacterium* sequence' (Crubézy et al., 1998). Subsequent work with this DNA suggests that it was an ancestral or archaic form of *Mycobacterium tuberculosis*, which existed at the time when urban life emerged in Egypt beginning around 3400 BC (Crubézy et al., 2006). The example of the skeleton of another young child with multiple bone tuberculosis from the Predynastic cemetery of Adaima published in 2011 'provides a picture of a period where tuberculosis must have been endemic throughout the population during the origins of urban settlement' (Dabernat and Crubézy, 2011).

The biomolecular work of Zink and Nerlich and their team continued. DNA was extracted from bone and tissues samples from Early Dynastic period burials at Abydos, and burials from the West Bank of Thebes, dating from the Middle Kingdom to Second Intermediate Period, and from the New Kingdom to the Late Period. First thirty-seven skeletal tissue samples were tested (Zink et al., 2001), and then the number grew to eighty-three. Out of eighty-three samples, eighteen of them tested positive for *Mycobacterium tuberculosis* complex DNA. Six positive results came from individuals with macroscopic evidence of 'tuberculous spondylitis', five from individuals with 'non-specific pathological alterations', and seven from individuals with 'normally appearing vertebral bones' (Zink et al., 2003, pp.242-244). The team concluded that for about 2500 years the frequency of tubercular disease remained the same, and that 'this is the first evidence for an extensive presence of tuberculosis in various ancient Egyptian populations' (Zink et al., 2003, p.248). Two years later the team had analysed a total of one hundred and sixty bone and tissue samples, and thirty-eight of the samples, coming from all three of the different time periods, 'tested positive for the presence of mycobacterial DNA' (Zink, Köhler, and Motamedi, 2005, p.85).

These samples were further characterized by spoligotyping, and all showed a *Mycobacterium tuberculosis* or a probable *Mycobacterium africanum* signature (Zink et al., 2003, p.365; Zink et al., 2004, p.411). The Early Dynastic material produced evidence for an ancestral strain of *Mycobacterium tuberculosis*, while the Middle Kingdom samples were characterized by *Mycobacterium africanum* strains. The samples from the New Kingdom to the Late Period revealed 'a modern strain of *M. tuberculosis*, but not the ancestral strain seen in the Pre-to Early Dynastic period' (Zink et al., 2007, p. 388). No evidence for the strain of *Mycobacterium bovis* was found in any of these samples. It has generally been thought that *M. bovis* was ancestral to *M. tuberculosis*, and that tuberculosis passed to humans at the time they domesticated cattle, which would have been approximately 4500-5000 BC in Egypt (Roberts and Manchester, 2005, p.184; Nerlich and Lösch, 2009). A more recent hypothesis is that human *M. tuberculosis* is the most ancient strain, and ancestral to *M. bovis (Brosch, et al., 2002)*

Lipid signatures

One last development in the search to identify ancient tuberculosis is the technique to detect characteristic lipid components, or mycolic acids in the *Mycobacterium tuberculosis* cell wall. Mycolic acid can be extracted from a bacterial culture and examined by HPLC, or high performance liquid chromatography, which gives 'a characteristic trace that identifies the species of origin' (Brown and Brown, 2011, p.872). This technique was recently used in the re-examination of the Granville mummy, a Late Period older female named Irtyersenu from Thebes. This mummy had been first autopsied by Dr. Granville in 1825, who declared that ovarian cancer had been the cause of her death; recent studies state that this tumour was a benign cystadenoma (Donoghue et al., 2010, p.51). Since a histological study in 1994 noted a pulmonary exudate, samples from her lungs, gall bladder and membranous tissues were tested for *Mycobacterium tuberculosis* DNA, and samples from her femurs and lung were tested by HPLC for mycolic acids of *Mycobacterium tuberculosis*. All DNA and mycolic acid samples were positive, and it would appear that an active tuberculosis infection was the cause of Irtyersenu's death (Donoghue, 2010).

Conclusions

Our understanding of the nature and extent of tubercular disease in ancient Egypt has changed dramatically in the last century. Based on both skeletal and DNA evidence from the Upper Egyptian site of Adaima, Crubezy has been able to demonstrate that tuberculosis was 'endemic' at the time of urban settlement in early Egypt. This conclusion has been expanded upon by Nerlich and Zink who have retrieved DNA from skeletal samples from both Abydos and Thebes dating to later Pharaonic periods. They have not only shown that tuberculosis was widespread in the ancient Egyptian population over a period of about 2, 500 years, but they have identified that different strains of Mycobacterium were present at different time periods. The latest development in the identification of tuberculosis in ancient human remains has been to detect mycolic acid in *Mycobacterium tuberculosis* cell walls. This makes the identification of tuberculosis possible, even if its DNA cannot be successfully retrieved.

References

Brosch, R., Gordon, S.V., Marmiesse, M., Brodin, P., Buchrieser, C., Eigelmeier, K., Garnier, T., Gutierrez, C., Hewinson, G., Kremer, K., Parsons, I. M., Pyn, A.S., Samper, S., van Soolingen, D., Cole, S.T., 2002. A new evolutionary scenario for the *Mycobacterium tuberculosis* complex. *Proceedings of the National Academy of Science of the United States of America,* 99 (6), pp.3684-3689.

Brown, T. and Brown, K., 2011. *Biomolecular Archaeology: An Introduction.* Oxford: Wiley-Blackwell.

Crubézy, É., Legal, L., Fabas, G., Dabernat, H. and Ludes, B., 2006. Pathogeny of archaic mycobacteria at the emergence of urban life in Egypt (3400 BC). *Infection, Genetics and Evolution,* 6, pp.13-21.

Crubézy, É, Ludes, B., Poveda, J.D., Clayton, J., Crouau-Roy, B. and Montagnon, D., 1998. Identification of *Mycobacterium* DNA in an Egyptian Pott's disease of 5,400 years old. *Comptes- rendus de l' Académie des sciences-Series III-Science de la vie,* 321, pp.941-951.

Dabernat, H. and Crubézy, É., 2011. Multiple Bone Tuberculosis in a Child from Predynastic Upper Egypt (3200 BC). *International Journal of Osteoarchaeology,* 20 (6), pp.719-730.

Derry, D., 1909. Anatomical Report. In *The Archaeological Survey of Nubia; Bulletin No. 3,* National Printing Department, Cairo, pp.29-52.

Donoghue, H, Lee, O., Minnikin, D., Gurdyal, S.B., Taylor, J.H. and Spigelman, M., 2010. Tuberculosis in Dr. Granville's mummy: a molecular re-examination of the earliest known Egyptian mummy to be scientifically examined and given a medical diagnosis. *Proceedings of the Royal Society-B Biological Sciences,* 277, pp.51-56.

Morse, D., Brothwell, D., and Ucko, P., 1964. Tuberculosis in Ancient Egypt. *The American Review of Respiratory Diseases,* 90, pp.524-541.

Nerlich, A., Haas, C., Zink, A., Szeimies, U. and Hagedorn, H., 1997. Molecular evidence for tuberculosis in an ancient Egyptian mummy. *The Lancet*, 350, p.1404.

Nerlich, A. and Lösch, S., 2009. Paleopathology of Human Tuberculosis and the Potential Role of Climate. *Interdisciplinary Perspectives on Infectious Disease*, 2009, Article ID 437187 9 pages.

Paabo, S., 1985. Molecular cloning of Ancient Egyptian mummy DNA. *Nature*, 314, pp.644-645.

Roberts, C. and Manchester, K., 2005. *The Archaeology of Disease*. Third edition. New York: Cornell University Press.

Ruffer, M.A., 1921. Pottsche Krankheit an einer Ägyptischen Mumie aus der Zeit der 21. Dynastie. In: *Studies in Paleopathology of Egypt*. Chicago: University of Chicago Press, pp.3-10.

Strouhal, E., 1991. Vertebral tuberculosis in ancient Egypt and Nubia. In Ortner, D.J. and Aufderheide A.C., eds *Human Paleopathology: Current Syntheses and Future Options*. Washington DC: Smithsonian Institution, pp.181-194.

Strouhal, E., 1999. Ancient Egypt and tuberculosis. In Pálfi, G.Y., Dutour, O., Deák, J. and Hutás, I., eds *Tuberculosis: past and present*. Budapest: Golden Book Publisher, pp.453-460.

Zimmerman, M., 1979. Pulmonary and Osseous Tuberculosis in an Egyptian Mummy. *Bulletin of the New York Academy of Medicine,* 55, pp.604-608.

Zink, A.R., Grabner, W., Reischl, U., Wolf, H. and Nerlich, A.G., 2003. Molecular study on human tuberculosis in three geographically distinct and time delineated populations from ancient Egypt. *Epidemiology and Infection*, 130, pp.239-249.

Zink, A., Haas, C.J., Reischl, U., Szeimies, U., and Nerlich, A.G., 2001. Molecular analysis of skeletal tuberculosis in an ancient Egyptian population. *Journal of Medical Microbiology*, 50, pp.355-366.

Zink, A.R., Hagedorn, C.J., Szeimies, H. and Nerlich, A., 1999. Morphological and molecular evidence for pulmonary and osseous tuberculosis in a male Egyptian mummy. In: Pálfi, G.Y., Dutour, O., Deák, J. and Hutás, I., eds *Tuberculosis: Past and Present*, Budapest: Golden Book Publisher, pp.379-383.

Zink, A.R., Molnár, E., Motamedi, N., Pálfy, G., Marcsik, A. and Nerlich, A.G. 2007. Molecular History of Tuberculosis from Ancient Mummies and Skeletons. *International Journal of Osteoarchaeology*, 17, pp.380-391.

Zink, A.R., Köhler, S., Motamedi, N., Reischl, U., Wolf, H. and Nerlich, A.G., 2005. Preservation and Identification of ancient M. tuberculosis complex DNA in Egyptian mummies. *Journal of Biological Research*, 80 (1), pp.84-87.

Zink, A.R., Sola, C., Reischl, U., Grabner, W., Rastogi, N., Wolf, H. and Nerlich, A.G., 2003. Characterization of *Mycobacterium tuberculosis* Complex DNAs from Egyptian Mummies by Spoligotyping. *Journal of Clinical Microbiology*, 41 (1), pp.359-367.

Zink, A.R., Sola, C., Reischl, U., Grabner, W., Rastogi, N., Wolf, H. and Nerlich, A.G., 2004. Molecular Identification and Characterization of *Mycobacterium tuberculosis* Complex in Ancient Egyptian Mummies. *International Journal of Osteoarchaeology*, 14, pp.404-413.

Palaeopathology, disability and bodily impairments

Sonia Zakrzewski

Archaeology, University of Southampton, Southampton, UK

Abstract

In archaeology, disabled people and disability have often been overlooked or considered 'hidden from view' (see Waldron, 2000). Yet disease and disability are present in all societies, and any person may become disabled at some point in their life. This disability may be permanent or temporary, and may contribute to social exclusion and the concept of 'difference'. 'What is perceived as a 'disability' or as 'madness' in one society, in another may be considered as just one attribute among many which make up an individual, or may not be perceived as part of the individual at all' (Waldron, 2000, p.7). Although Egypt seems to have been relatively accepting towards individuals considered as 'different' or 'other' (Jeffreys and Tait, 2000), Egyptian attitudes towards minorities (of any form, be they physical or ethnic) are varied.

Theories of disability

According to the World Health Organisation (WHO), disability is an umbrella term, covering impairment, activity limitation, and restriction on participation (WHO, no date). An impairment is a problem in body function or structure. An activity limitation is a difficulty encountered by an individual in executing or undertaking a task or action. A participation restriction is a problem experienced by a person in their involvement in life situations. Disability is therefore a complex phenomenon, reflecting an interaction between the features of a person's body and the features of the society in which they live.

Disability, however, is constructed in different ways by different academic disciplines. Whereas the medical community have generally considered disability in terms of medical reductionism, social scientists have, more commonly, followed a social model of disability (Thomas, 2007). Oliver (1983) argued that disability is not caused by impairment, but rather from social restrictions placed upon individuals with bodily impairment. Resultingly, medical sociologists theorise chronic illness and disability through a lens of social deviance (Thomas, 2007). This focus comprises both aspects of the impaired body and the lived experience. In this sense, a disability is simply a form of limited activity, and hence a disabled person is one who has a medically certifiable condition preventing them from carrying out the full 'normal' range of age-related activities (Thomas, 2007). Following this argument, disability is an age-related and universal phenomenon, with the importance being placed upon living with 'illness' and hence the focus not being constrained to the individual, but also to the changed circumstances of significant others.

In contrast, Murphy (1990), despite rejecting the social deviance model, views disability as a state of social liminality, whereby the individual is in a state of exclusion from ordinary life and is denied the full expression of 'being human'. As such, the disabled person is outside the formal social system. This construction of disability is built upon human perception and 'being' as embodied phenomena, with meaning residing *in* the body and the body itself residing *in* the world (Merleau-Ponty, 1962). Consequently, there is a fluid boundary between disabled and able-bodied, with identity (and particularly self-identity and ascribed identity) being of significance. Negative views of disability, or disablism, develop this viewpoint and stress that the relative degrees of need, care or dependency, may lead to individuals being ascribed a childlike status. The varying configurations of needs are thus given primacy at the expense of the social individual, thereby revolving back to the social model of disability (Thomas, 2007).

Medical sociologists appear to congregate around the view that disability and impairment should not be viewed in terms of biological reductionism, but rather that disability and being disabled are *not* all about the body, but rather comprise impairment effects and hence are subjective experiences with many differences (Thomas, 2007). Tremain (2002) develops this argument further to consider impairment and disability to be viewed as sex is to gender. Thus the impaired body is socially constructed, and the embodied 'difference' is the so-called impairment and external reaction to this is the so-called disability. This permits 'disabled' to be viewed as a point upon a continuum rather than as a binary opposition to 'able-bodied'. Consequently, focus is placed upon the body in pain, chronic illness such as rheumatoid arthritis, and other forms of illness, including mental illness (Thomas, 2007).

This paper follows all these arguments to explore what is meant by disability within an explicitly Egyptian context. Furthermore, it evaluates the impact that impairment or difference, relative to the Egyptian norm, might have had upon lived social experience. It is clear from texts, such as the 'Instruction of Amenemope' (which includes phrases such as "Do not laugh at a blind man, Nor tease a dwarf, Nor cause hardship for the lame"), that tolerance towards people with disabilities was recommended. This has been described as 'a more generous attitude towards some disabilities' (Quarmby, 2011, p.25). The above concepts of disease and disability link in to constructions of 'otherness' (Hubert, 2000) and identity within bioarchaeology (Knudson and Stojanowski, 2008; Perry, 2007). However, distinctions remain in terms of *which* disabilities or 'others' are deemed respectable or viewed as 'Egyptian people'.

Representations of disablement and malformation

Within Egyptian artistic representation, the body is viewed as an entity, but with each portion having an idealised or typical form which are then combined to form a composite body (Robins, 1994). Despite this leading to a human figure that plainly does not correspond directly with reality, some representations do show the body in meticulous anatomical detail (Weeks, 1970). As a result, most individuals are depicted in an idealised form. However, some specific individuals were depicted differently from other 'normal' individuals, for example dwarves (Dasen, 1993; Iversen, 1975; Robins, 1994; Weeks, 1970), but physical irregularities or impairments were primarily shown for minor figures (Dasen, 1993).

Probably the most common representation of disability is of dwarfing. This phrasing could be construed as an oxymoron, as small stature and dwarfing need not lead to any reduction in ability to undertake activities. Indeed, there were even three distinct Egyptian words for abnormally short people (*dng*, *nmw*, and *ḥwꜥ*), and use of these words was usually accompanied by a determinative depicting a disproportionate dwarf with a long trunk and short limbs (Dasen, 1993). Several dwarves are well known and were of high social ranking. During the Fourth Dynasty (although previously dated as Fifth or Sixth Dynasty (e.g. Filer, 1995; Kozma, 2006)), the dwarf Perniankhu, a court entertainer, was buried in the great western cemetery at Giza (Wilkinson, 2007). In statues he is depicted with short bowed legs, thick ankles and flat feet. However, he is also depicted with symbols of authority, such as a sceptre and long staff (Hawass, 1991; Wilkinson, 2007). Also during the Old Kingdom, the dwarf Seneb achieved the rank of court official, as priest for Khufu and Djedefra and 'Director of Dwarfs in Charge of Dressing' [the sovereign], and tutor to the king's son (Wilkinson, 2007). Seneb was buried with his wife in a mastaba at Giza, and is well known from statuary, such as the sculptural group with his family housed in the Cairo Museum [Cairo Museum JE 51280]. These relatively anatomically accurate depictions of achondroplastic dwarfism continue through to the Late Period, as evidenced by illustration of the naked figure of Djeho on his sarcophagus [Cairo Museum CG 29307 (his patron was Tjaiharpta CG 29306)] (Baines, 1992; Kozma, 2006). The 'physical anomaly [of dwarfing] was not only tolerated, but accepted and valued as a divine mark for its religious associations' (Dasen, 1993, p.156), as not only do dwarves appear to have a special affinity with solar deities, but their use in cult dances is referred to in the Pyramid Texts (Baines, 1992).

Dwarves are thus depicted in an essentially positive manner (Sullivan, 2001), and their biological disorder appears not to have been viewed within artistic representation as a handicap or disability. It has been argued, however, that they may have been considered as liminal, due to their frequent association with other malformed people (such hunchbacks) and exotic people (such as from Nubia or Punt) (Dasen, 1993).

In this sense, physical attributes might be considered of importance in depiction of personhood. Although individual and personal traits were generally avoided (Iversen, 1975), and bodies were depicted in significant meaningful detail, such as legs with muscles flexed (Weeks, 1970), representations exist that might show some form of disease process or disability. Depictions of individuals with hunched or humped backs are relatively common. Examples include the gardener from the tomb of Ipuy at Beni Hasan [Metropolitan Museum of Art, New York 30.4.115], two Predynastic wooden figurines curated in Brussels [Musées Royaux d'Art et d'Histoire de Bruxelles], or the well-known Predynastic red clay statuette from Aswan [Private collection, Paris]. The deformities illustrated by these examples are commonly thought to be representations of kyphosis resulting from Pott's disease (Filer, 1995; Halioua and Ziskind, 2005; Reeves, 1992; Ziskind and Halioua, 2007), although other putative causes have been suggested (Nunn, 1996). Both congenital deformity (*talipes equinus*) and bodily change as a result of infection (poliomyelitis) have been suggested for the withered right leg depicted on the New Kingdom funerary stela of Roma [Ny Carlsberg Glyptotek, Copenhagen AIN 134] (Filer, 1995; Halioua and Ziskind, 2005; Nunn, 1996). This stela is of particular note as the deceased is shown using his staff as a crutch rather than as a symbol of status and rank (Jeffreys and Tait, 2000).

Focus has heretofore concentrated upon representations of skeletal manifestations of disability. It should be remembered, however, there are many other forms of disablement, such as reductions or impairments in hearing and vision. The blind appear most commonly depicted in specific roles, such as harpists or singers, although in some iconographic representations blindness has been suggested to represent piety (Dasen, 1993). Examples include the blind harpist from the New Kingdom tomb of Nakht at Thebes [Tomb TT52, Metropolitan Museum, New York 15.5.19d], the blind harpist from the New Kingdom tomb of Patenemheb at Saqqara [Rijksmuseum van Oudheden, Leiden AMT 1-35], and Raia. The latter was Chief of Singers in the temple of Ptah at Memphis during the Ramesside period (Wilkinson, 2007), and, in his tomb at Saqqara, was depicted as blind when playing music for his patron deities, Ptah and Hathor, but as sighted in other scenes (Dasen, 1993).

In addition to these examples of disability, note must be made of iconographic representations of disease, such as the genital hypertrophy suggestive of schistosomiasis or the distended abdomens suggestive of umbilical hernias (Jeffreys and Tait, 2000), such as in the Old Kingdom tomb of Ptah-Hetep at Saqqara (Thompson Rowling, 1967a). Furthermore the medical papyri provide evidence of treatment of these conditions, such as the twenty treatments given in the Ebers papyrus for haematuria (Thompson Rowling, 1967b). In modern Egyptian medicine, this is commonly associated with schistosomiasis or neoplasia of the renal tract (Thompson Rowling, 1967b). Although these disorders are not commonly considered as disabilities, these can be disabling to the individuals and society concerned as they can to reductions in fitness, activity or participation, hence falling under the WHO classification for disability.

Palaeopathologies of Egyptian disablement and impairment

Given the skeletal and mummified preservation of bodies from Egyptian contexts, there are relatively large numbers of individuals who might be considered, in palaeopathological terms, to have experienced potential disablement or impairment. Only a few examples can be included here, and hence should be considered as case studies for discussion.

As noted earlier, the most commonly depicted congenital skeletal anomaly is that of dwarfing. Skeletal examples of putative achondroplasia include the two dwarfs found in chambers M and L of the Early Dynastic tomb of Semerkhet [Natural History Museum, London AF11.4/427 & AF 11.4.462] (Dasen, 1993; Ortner and Putschar, 1981) and Predynastic long bones that may derive from El-Mostagedda (Brothwell, 1967; Dasen, 1993). The latter are potentially Badarian in date and may be curated in the Duckworth Collection, Cambridge. More recently, the skeleton of Perniankhu was excavated from tomb G1700 at Giza (Hawass, 1991). Brief descriptions are to be found in Filer (1995) and Kozma (2006), although a full palaeopathological description of the skeleton has yet to be published. All skeletal remains of dwarfs recorded in the literature (including those whose whereabouts are now unknown) appear to have been relatively 'healthy' and may have been considered 'other' rather than 'disabled'.

Of particular note among other congenital disorders, diseases and/or malformations, described from Egyptian skeletal and mummified remains, is the suggestion of cerebral palsy for a mature 13th Dynasty female from Thebes (Nerlich et al., 2010). The authors note strong flexion in the radio-carpal joint and hyperextension in the metacarpo-phalangeal joints of the left hand, associated with greater dental wear & abrasion of the temporo-mandibular joint on the left side. They argue that this could arise from compensatory use of the left side of jaw as, in addition, they note some evidence of reduced functionality of the right side masticatory musculature. People with cerebral palsy may also suffer from dyskinesia and ataxia, and hence can experience the side effect of drooling and difficulty with speech associated with 'strange' facial expressions (Nerlich et al., 2010). The most common skeletal malformation is likely *spina bifida occulta*. Although usually asymptomatic, spina bifida occulta may cause pain and neuralgia. One Nubian individual [RCS Nubian 178A, Natural History Museum, London] has even been argued to have been paraplegic, associated with spina bifida (Thompson Rowling, 1967c).

Most 'disabilities' arise during the lifetime of the individual concerned, but may not be permanent effects. For example, limitations upon participation may occur either permanently or temporarily as a result of bodily trauma. Post-traumatic ankylosis of bones, such as the talocrural joint of male E12 from the Memphite tomb of Horemheb IV (Strouhal, 2008), might affect mobility. As a result of a femoral fracture, one Egyptian has been described as '[having] had one leg slightly shorter than the other and walked with a limp' (Filer, 1995, pp. 87-88). Given the illustrations, Filer is most likely referring to individual EA 37340 at the British Museum, London. Filer (1995, p.88) continues, 'the long period of immobility whilst the leg set would have made the person dependant [sic] upon his community'. The most problematic issue with trauma is the mal-alignment of fractures. The medical papyri, including the Edwin Smith papyrus, indicate that the fractures were reduced through manipulation and the use of splints. Archaeological evidence of the latter are known, such as the palm tree splints supporting two Christian period Nubian arms which are curated at the Hunterian Museum in London. Poorly aligned ulnae have been argued to be the most common of poorly aligned fractures (Hussien et al., 2010), and thus would have affected productivity and activity.

Iconographic evidence for Pott's disease has been summarised earlier. Tuberculosis has been recognised palaeopathologically on the basis of both skeletal kyphosis and iliopsoas abscesses, such as in the 21st Dynasty Theban mummy Nespaheran (Filer, 1995; Nunn, 1996; Ruffer, 1921; Ziskind and Halioua, 2007). For reviews of such cases see Buikstra et al. (1993) and Morse et al. (1964). Mycobacterial DNA has been recovered from a variety of Egyptian contexts (such as from Dr Granville's mummy, a 26th Dynasty female named Irtyersenu (Donoghue et al., 2010)), although not all DNA fragments have been sufficiently well preserved to distinguish between *M. bovis*, *M. africanum* and *M. tuberculosis*, such as from an Adaïma child (Crubézy et al., 1998). Given the successful amplification of *M. tuberculosis* from individuals not exhibiting skeletal changes associated with tuberculosis (such as Dr Granville's mummy mentioned above), it has been argued that infection with *M. tuberculosis* was common (Zink et al., 2001; 2003) or even endemic (Donoghue et al., 2010).

Acquired infections and parasitic conditions such as schistosomiasis are also recognisable, e.g. *Schistosoma* ova found in the intestinal tract of Nakht (an unembalmed New Kingdom mummy, sometimes denoted as Royal Ontario Museum I or ROM I) (Millet et al., 1980). Schistosome circulating anodic antigen has been identified in X-group mummies from Wadi Halfa (Miller et al., 1992) and in Predynastic and New Kingdom mummies (Deelder et al 1990). Other parasites such as *Ascaris* (Cockburn et al., 1980) [as in PUM II, belonging to Philadelphia Art Museum, but curated at Philadelphia University Museum], filarial worms [as in the Leeds mummy Natsef Amun] (Sandison and Tapp, 1980) or guinea worms [as in Manchester mummy 1770] (Sandison and Tapp, 1980) have been described. Lymphatic filariasis, as occurs when filarial worms block lymphatic channels, has been proposed as the cause of elephantiasis noted in some iconographic representations (Sandison and Tapp, 1980; Weeks, 1970). Such infections could cause declines in health and sense of wellbeing, and potentially in activity levels.

Some idiopathic disorders, such as scoliosis, are relatively easily identified palaeopathologically, such as burial 11 from Quesna (Rowland, 2008). Mild forms are unlikely to have any major effect upon the person affected, but more severe forms are associated with thoracic insufficiency and hence difficulties in breathing (Campbell et al., 2003). Although no examples of congenital deafness have been found, hearing loss or impairment has been argued for a mature male individual from Giza who suffered from a slice injury to the left temporal and into the auditory meatus (Filer, 1992). This can be compared with congenital deafness resulting from bone growth over the external auditory meatus described in a Roman child [Poundbury 1114] in the UK (Molleson, 1993). Finally, although sight is difficult to assess palaeopathologically, a possible case of blindness deriving from carcinoma has been noted from Naga-ed-Dêr (Podzorski, 1990). For a review of eye diseases in Egypt, see Andersen (1997).

Integration of dis/ability within palaeopathology

Roberts (1999) enumerates difficulties in considering disability within a palaeopathological framework. These issues include aspects such as modern concepts of disability differing from ancient views, difficulties in assessing what abnormalities in skeletal record may have been disabling to the individual concerned, difficulties in assessing the impact upon the society in which that individual lived, problems in interpreting evidence from artistic or documentary sources to get a degree of prevalence within the population, and issues with the archaeological record for caring and compassion. Dettwyler (1991) argued that conclusions cannot be drawn about the quality of life for the disabled from skeletal evidence of impairment, but if bioarchaeologists move beyond implicit (and often tacitly negative) assumptions about disability, focus can be placed upon both the individuals as people and upon 'ability'. This argument develops from Dettwyler's demonstration that 'survival' of individuals with disabilities or impairments cannot be directly assumed to result from compassion. Here I am arguing that disabilities or impairments should be viewed, following Tremain (2002), as simply points upon a continuum of ability, and that the recognition of these positions may be elucidated by other aspects of Egyptian archaeology. In this sense, bioarchaeologists need to remember that it is bodily impairments, such as joint disease or arthritis, rather than disability, that are excavated and studied (Cross, 1999), and subsequently a concept of disability, based upon a more explicitly social model, must be developed.

Using skeletal manifestations of potentially disabling conditions permits a baseline model of physical ability to be developed. It is important to note that this is only a baseline, as both bodily and social adaptation may have been used by the individual and their society to develop their abilities (or specialisations, for example, as musicians). In the past, pain and dependency too frequently have been assumed to have been experienced by the individuals concerned (see discussion in Roberts, 2000). There are, however, no standards of osseous or other bodily change that can be demonstrated to cause pain, or indeed the degree of pain experienced. Relatively minor bone damage resulting from osteoarthritis may produce severe pain (Roberts, 1999), and yet greater skeletal changes may be asymptomatic. Furthermore, it is well-known that bodily deformity, such as resulting from leprosy, need not necessarily be disabling (Roberts, 1999).

These aspects demonstrate the importance of developing a social model of ability from the palaeopathological record for impairment. Using this framework, the degree of disablement can be considered. Health problems such as anaemia, although not particularly 'disabling', also affect the individual's life, such as through increased fatigue, loss of body weight and effects upon their sense of health. These in turn may affect productivity and thus their 'ability'. Such individuals might not be disabled under the common usage of the term, but would experience 'restriction upon participation' and thus might be considered as being in some group of 'others' by the surrounding society.

It has already been noted that iconographic representation of some 'disabled' groups within Egypt indicates that such people 'were not regarded as subnormal or in any way socially diminished' (Jeffreys and Tait, 2000, p.91). As these authors argue, the medical conditions included, e.g. dwarfing, blindness, those with clubfeet etc., are those that are incurable and are primarily congenital. By contrast, they argue that conditions developing during life, e.g. spinal kyphosis resulting from tuberculosis, were depicted in much more negative tones or were considered 'emblematic of non-elite rank and activities' (Jeffreys and Tait, 2000, p.92). Consequently, using the term disability in ancient Egypt is almost meaningless unless the medical conditions being considered are suitably quantified. Furthermore, palaeopathologists and bioarchaeologists also need to consider that conditions not deemed disabling in modern conditions, such as myopia, may have had a much greater impact upon *both* the individual and the society.

Developing an Egyptian social model of disability and impairment

The arguments above demonstrate that disability should be considered as a normal part of life, and that all individuals can be placed upon a fluid continuum of ability. During certain periods of life, an individual might be less physically 'able' or be restricted in their participation, such as during pregnancy or when recuperating following skeletal trauma or fracture. For example, during the healing of a bone following fracture, it must be immobilised, thereby reducing mobility and the use of that bone or an affected limb. It is possible that the Egyptians recognised this as Sullivan (2001, p.262) argues that '[the Egyptian] artist does not attempt to 'beautify' deformity which suggests that there was a prevailing attitude of cultural acceptance over deformity'. She links this to physical deformities of certain Egyptian deities, such as Bes and Hapi, and suggests that 'physical deformity may have been received as a positive mark of divinity' (Sullivan, 2001, p.262). There is little evidence of prejudice, at least for disorders that might be considered 'noble' (Jeffreys and Tait, 2000, p.91), and hence temporal duration, context and social rank must be incorporated into all Egyptian models of disability and impairment.

Furthermore, although chronic illnesses and impairments appear in all sectors of modern society, certain groups, such as the lower ranking or poorer, are more vulnerable (Bartley, 2004). Following this hypothesis for past Egyptian populations, it is imperative that focus is placed upon the labourers, farmers and lower ranking individuals in order to develop a model of Egyptian health and ability. An 'anthropologie de terrain' approach (Duday, 2009) might permit small-scale differences in funerary practice or relative liminality (cf Knüsel, 1999), as might happen within primarily lower ranking groups, to be identified (such as the prone burial of congenitally deaf child at Poundbury (Molleson, 1999)). Furthermore, given that 'the personal responses of disabled individuals to their impairments ... have to be located within a contextual framework that takes account of both history and ideology' (Oliver and Barnes, 2012, p.98), the ability continuum must be viewed through the lens of the individual agent and the moment in their life course. Ancient Egyptians may have viewed individuals along some form of ability continuum, with people acting differently in differing situations and in different periods of their lives. As archaeologists and palaeopathologists, we need to move beyond a negative framework for impairment, but rather view all people in terms of varying abilities, some of which may have potential for exploitation. In addition to the blind harpists discussed above, Nunn (1996) has suggested that individuals with myopia (who might be considered blind or partially sighted in some contexts) might have been employed to undertake miniature engravings. Indeed, in modern western society, some people born deaf have argued that, due to use of sign language, they should not be considered disabled, but rather a cultural minority (Oliver and Barnes, 2012). A social model of Egyptian 'disability' would argue that all people have varying abilities, rather than disabilities and impairments. Dis/ability therefore acts as one dimension within Egyptian identity. Thus to access detailed aspects of the past Egyptian attitudes to dis/ability, all cadavers need to be remembered to have been people, rather than categorised as disabled or 'other', and importance placed upon the timing and duration of all bodily effects.

Acknowledgments

The author would like to thank Stephanie Wright (for discussions regarding disability in ancient Rome), Joanne Rowland, Sarah Inskip, Scott Haddow and all the others in the Minufiyeh (Quesna) team, and the organisers of the Palaeopathology in Egypt and Nubia workshop, and most especially Ryan Metcalfe. Partially funded by the Egypt Exploration Society and the Arts & Humanities Research Council.

References

Andersen, S.R., 1997. The eye and its diseases in ancient Egypt. *Acta Ophthalmologica Scandinavica* 75, pp.338-344.

Baines, J., 1992. Merit by Proxy: The Biographies of the Dwarf Djeho and his patron Tjaiharpta. *Journal of Egyptian Archaeology* 78, pp.241-257.

Brothwell, D., 1967. Major congenital anomalies of the skeleton: Evidence from earlier populations. In: D. Brothwell and A.T. Sandison, eds. *Diseases in Antiquity*. Springfield: Charles C Thomas. pp.423-443.

Buikstra, J.E., Baker, B.J. and Cook, D.C. 1993. What diseases plagued the ancient Egyptians? A century of controversy considered. In: W.V. Davies and R. Walker, eds. *Biological Anthropology and the Study of Ancient Egypt*. London: British Museum Press, pp.24-53.

Campbell, R.M., Smith, M.D., Mayes, T.C., Mangos, J.A., Willey-Courand, D.B., Kose, N., Pinero, R.F., Alder, M.E., Duong, H.L. and Surber, J.L. 2003. The characteristics of thoracic insufficiency syndrome associated with fused ribs and congenital scoliosis. *Journal of Bone and Joint Surgery of America* 85, pp.399-408.

Cockburn, A., Barraco, R.A., Peck, W.H and Reyman, T.A. 1980. A classic mummy: PUM II. In: A. Cockburn, E. Cockburn and T.A. Reyman, eds. *Mummies, Disease and Ancient Cultures*. Cambridge: Cambridge University Press. pp.69-90.

Cross, M. 1999. Accessing the Inaccessible: Disability and archaeology. *Archaeological Review from Cambridge* 15, pp.7-30.

Crubézy, É., Ludes, B., Poveda, J-D., Clayton, J., Crouau-Roy, B. and Montagnon, D., 1998. Identification of Mycobacterium DNA in an Egyptian Pott's disease of 5400 years old. *Comptes Rendus de l'Académie de Sciences de Paris, Sciences de la vie* 321, pp.941-951.

Dasen, V., 1993. *Dwarfs in Ancient Egypt and Greece*. Oxford Monographs in Classical Archaeology. Oxford: Clarendon Press.

Deelder, A.M., Miller, R.L., DeJonge, N. and Kruger, F.W., 1990. Detection of schistosome antigen in mummies. *The Lancet* 335, pp.724-725.

Dettwyler, K.A., 1991. Can paleopathology provide evidence of 'compassion'? *American Journal of Physical Anthropology* 84, pp.375-384.

Donoghue, H.D., Lee, OY-C., Minnikin, D.E., Besra, G.S., Taylor, J.H. and Spigelman, M., 2010. Tuberculosis in Dr Granville's mummy: A molecular re-examination of the earliest known Egyptian mummy to be scientifically examined and given a medical diagnosis. *Proceedings of the Royal Society B* 277, pp.51-56.

Duday, H., 2009. *The Archaeology of the Dead: Lectures in Archaeothanatology*. Oxford: Oxbow.

Filer, J.M., 1992. Head injuries in Egypt and Nubia: A comparison of skulls from Giza and Kerma. *Journal of Egyptian Archaeology* 78, pp.281-285.

Filer, J., 1995. *Egyptian Bookshelf: Disease*. London: British Museum Press.

Halioua, B. and Ziskind, B., 2005. *Medicine in the Days of the Pharaohs*. Transl. MB DeBevoise. Cambridge: Harvard University Press.

Hawass Z 1991. The Statue of the Dwarf Pr-n(j)-'nḫ(w) Recently Discovered at Giza. *Mitteilungen des Deutschen Archäologischen Instituts Abteiling Kairo*, 47, pp.157-162.

Hubert J 2000. Introduction: the complexity of boundedness and exclusion. In: Hubert J. ed. Madness, Disability and Social Exclusion. *One World Archaeology*, 40. London: Routledge. pp.1-8.

Hussien, F., El Banna, R., Kandeel, W. and Sarry El Din, A., 2010. Similarity of Fracture Treatment of Workers and High Officials of the Pyramid Builders. In: J. Cockitt and R, David., eds. *Pharmacy and Medicine in Ancient Egypt*. BAR International Series 2141. Oxford: Archaeopress, pp.85-89

Iversen, E., 1975. *Canon and Proportions in Egyptian Art*. Warminster: Aris & Phillips.

Jeffreys, D. and Tait, J., 2000. Disability, madness, and social exclusion in Dynastic Egypt. In: J. Hubert, ed. Madness, *Disability and Social Exclusion. One World Archaeology 40*. London: Routledge, pp.87-95.

Knudson, K.J. and Stojanowsk, C.M., 2008. New Directions in Bioarchaeology: Recent contributions to the study of human social identities. *Journal of Archaeological Research* 16, pp.397-432.

Knüsel, C.J., 1999. Orthopaedic Disability: Some hard evidence. *Archaeological Review from Cambridge* 15, pp.31-53.

Kozma, C., 2006. Dwarfs in Ancient Egypt. *American Journal of Medical Genetics* 140A, pp.303-311.

Merleau-Ponty, M., 1962. *The Phenomenology of Perception*. London: Routledge.

Miller R.L., Armelagos, G.J., Ikram, S., DeJonge, N., Kruger, F.W. and Deelder, A.M. 1992. Palaeoepidemiology of schistosoma infection in mummies. *British Medical Journal*, 304, pp.555-556.

Millet, N.B., Hart, G.D., Reyman, T.A., Zimmerman, M.R. and Lewin, P.K., 1980. ROM I: mummification of the common people. In: A. Cockburn, E. Cockburn and T.A. Reyman, eds. *Mummies, Disease and Ancient Cultures*. Cambridge: Cambridge University Press. pp.91-105.

Molleson, T., 1993. The Human Remains. In: D.E. Farwell and T.I. Molleson, eds. *Poundbury: The Cemeteries*. Dorset Natural History and Archaeological Society Monograph 11, pp.142-214.

Molleson, T., 1999. Archaeological Evidence for Attitudes to Disability in the Past. *Archaeological Review from Cambridge* 15, pp.69-77.

Morse, D., Brothwell, D.R. and Ucko, P.J., 1964. *Tuberculosis in ancient Egypt*. American Review of Respiratory Diseases 90, pp.524-541.

Murphy, R.F., 1990. *The Body Silent*. London: Norton & Company.

Nerlich, A.G., Panzer, S., Hower-Tilmann, E. and Lösch, S., 2010. Palaeopathological – Radiological evidence for Cerebral Palsy in an Ancient Egyptian female mummy from a 13th Dynasty Tomb. In: J. Cockitt and R. David, eds. *Pharmacy and Medicine in Ancient Egypt*. BAR International Series 2141. Oxford: Archaeopress, pp.113-116.

Nunn, J.F., 1996. *Ancient Egyptian Medicine*. London: British Museum Press.

Oliver, M., 1983. *Social Work with Disabled People*. Basingstoke: Macmillan.

Oliver, M. and Barnes, C., 2012. *The New Politics of Disablement*. Basingstoke: Palgrave Macmillan.

Ortner, D. and Putschar, W., 1981. *Identification of Pathological Conditions in Human Skeletal Remains.* Smithsonian University Press: Washington.

Perry, M.A., 2007. Is Bioarchaeology a Handmaiden to History? Developing a Historical Bioarchaeology. *Journal of Anthropological Archaeology* 26, pp.486-515.

Podzorski, P.V., 1990. *Their Bones Shall Not Perish.* New Malden: SIA Publishing.

Quarmby, K. 2011. *Scapegoat: Why we are failing disabled people.* London: Portobello.

Reeves, C. 1992. *Egyptian Medicine.* Princes Risborough: Shire.

Roberts, C. 1999. Disability in the Skeletal Record: Assumptions, Problems and some examples. *Archaeological Review from Cambridge* 15, pp.79-97.

Roberts, C.A. 2000. Did they take sugar? The use of skeletal evidence in the study of disability in past populations. In: J. Hubert, ed. *Madness, Disability and Social Exclusion. One World Archaeology 40.* London: Routledge, pp.46-59.

Robins, G. 1994. *Proportion and Style in Ancient Egyptian Art.* Austin: University of Texas Press.

Rowland, J., 2008. The Ptolemaic-Roman Cemetery at the Quesna Archaeological Area. *Journal of Egyptian Archaeology* 94, pp.69-93.

Ruffer, M.A., 1921. *Studies in the Palaeopathology of Egypt.* Chicago: University of Chicago Press.

Sandison, A.T. and Tapp, E., 1980. Disease in Ancient Egypt. In: A. Cockburn, E. Cockburn and T.A. Reyman, eds. *Mummies, Disease and Ancient Cultures.* Cambridge: Cambridge University Press, pp.38-58.

Strouhal, E., 2008. *The Memphite Tomb of the Horemheb Commander-in-Chief of Tutankhamun IV. Human skeletal remains.* EES Excavation Memoir 87. London: Egypt Exploration Society.

Sullivan, R., 2001. Deformity: A modern western prejudice with ancient origins. *Proceedings of the Royal College of Physicians of Edinburgh* 31, pp.262-266.

Thomas, C., 2007. *Sociologies of Disability and Illness.* Basingstoke: Palgrave Macmillan.

Thompson Rowling, J., 1967a. Hernia in Egypt. In: D. Brothwell and A.T. Sandison, eds. *Diseases in Antiquity.* Springfield: Charles C Thomas, pp.444-446.

Thompson Rowling, J. 1967b. Urology in Egypt. In: D. Brothwell and A.T. Sandison, eds. *Diseases in Antiquity.* Springfield: Charles C Thomas, pp.532-537.

Thompson Rowling, J., 1967c. Paraplegia. In: D. Brothwell and A.T. Sandison, eds. *Diseases in Antiquity.* Springfield: Charles C Thomas, pp.272-278.

Tremain, S., 2002. On the Subject of Impairment. In M. Corker and T. Shakespeare, eds. *Disability/Postmodernity: Embodying disability theory.* London: Continuum. pp.32-47.

Waldron, T., 2000. Hidden or Overlooked? Where are the disadvantaged in the skeletal record? In: J. Hubert, ed. *Madness, Disability and Social Exclusion. One World Archaeology 40.* London: Routledge, pp.29-45.

Weeks, K.R., 1970. *The Anatomical Knowledge of the Ancient Egyptians and the Representation of the Human Figure in Egyptian Art.* Unpublished PhD Thesis: Yale University.

WHO, no date. *Disabilities.* [online] Available at:
<http://www.who.int/topics/disabilities/en/> [Accessed 20th August 2012].

Wilkinson, T., 2007. *Lives of the Ancient Egyptians.* London: Thames & Hudson.

Zink, A., Haas, C.J., Reischl, U., Szeimies, U. and Nerlich, A.G., 2001. Molecular analysis of skeletal tuberculosis in an ancient Egyptian population. *Journal of Medical Microbiology* 50, pp.355-366.

Zink, A.R., Grabner, W., Reischl, U. and Nerlich, A.G., 2003. Molecular study on human tuberculosis in three geographically distinct and time delineated populations from ancient Egypt. *Epidemiology & Infection* 130, pp.239-249.

Ziskind, B. and Halioua, B., 2007. La tuberculose en ancienne Égypte. *Revue des Maladies Respiratoires* 24, pp.1277-1283.

Dental infections in ancient Nubia

Roger J. Forshaw

KNH Centre for Biomedical Egyptology, The University of Manchester, Manchester, UK

Abstract

This paper is based upon the results of a dental survey of approximately 900 skulls which were excavated by the first Archaeological Survey of Nubia (1907-1911). These specimens are now located variously in in the KNH Centre for Biomedical Egyptology at Manchester University, the Duckworth Collection at the University of Cambridge and the Natural History Museum, London. The cemeteries from which the skulls were unearthed range in date from Predynastic through to the Christian Period.

The study examines the pathogenesis of dental infections and considers some severe examples of such infections, which may well have been the cause of fatalities in antiquity. A number of pathological bony cavities observed in the Nubian collections are highlighted and possible sequelae relating to causes of death are considered.

Introduction

Teeth are the hardest and most chemically stable tissues in the body and are highly resistant to decay in most burial environments. They form an important part of any bioarchaeological and palaeopathological investigations into the human remains of past populations. A study of dental disease provides information about the health and diet of an ancient society at a specific period in antiquity. The importance of good oral health should not be underestimated and the ability to masticate food well, particularly the tough abrasive food of ancient diets, is a significant factor in overall health status. Importantly, prior to the age of antibiotics, infections caused by neglected teeth could lead to life-threatening ailments and were almost certainly a cause of many deaths in antiquity.

This paper refers to a survey of teeth and their supporting structures, in a collection of Nubian skulls examined during 2011 and 2012 by the author. The study forms part of the 'Sir Grafton Elliot Smith and the Archaeological Survey of Nubia: their significance to the palaeopathological tradition' project sponsored by the Wellcome Trust. The aim of the odontological element of the survey is to record and analyse the frequencies of dental diseases in order to reveal more about the status of dental and indeed overall health in ancient Nubia. An online database has been created which will record this data and being open access is available for future research.

Materials and methods

The skulls utilised in the investigation are part of the skeletal remains excavated by the first Archaeological Survey of Nubia during the years 1907-1911. The skeletal material was unearthed from cemeteries adjacent to the river Nile in the region of Lower Nubia, which today is southern Egypt and northern Sudan. Much of this material has been dispersed to collections in various parts of the world, but approximately 900 skulls are today housed in the Natural History Museum London, the Duckworth Laboratory at the University of Cambridge and the KNH Centre at the University of Manchester. It is these collections that provided the sample material for this project.

All the samples were analysed for evidence of dental abnormalities in the following categories: carious lesions, ante-mortem tooth loss (AMTL), tooth wear and periapical lesions. These conditions were recorded by tooth and position within the dental arch as defined by Buikstra and Ubelaker (1994). Tooth wear was scored using the method as determined by Brothwell (1963). This paper discusses periapical lesions in alveolar bone and examines a number of these that were found within the skulls of the collections, with the full results of the overall study being the subject of a future publication.

Periapical bony cavities – discussion

Infection of the periapical tissues surrounding teeth can be as a result of tooth wear, caries, trauma, or as a consequence of periodontal disease. Both tooth wear and caries are related to diet and understanding a population's diet is an essential component in analysing dental health. Information concerning diet can be determined from archaeozoological, archaeobotanical and archaeological data (Gamza and Irish, 2010). Additionally, studies involving stable carbon and nitrogen isotope analysis can provide direct information about the average diet of an individual during life (Stenhouse and Baxter, 1979; DeNiro, 1987). Examples of such studies are Iacumin et al. (1998), Dupras et al. (2001) and Thompson et al. (2005).

The most frequent pathological condition identified in the study was that of excessive tooth wear, a condition that has been widely observed in investigations of ancient Nubian teeth (Ruffer, 1920; Leek, 1967; Hillson, 1979; Miller, 2008). This tooth wear increased with the age of an individual and varied from a slight polishing of the cusps to almost complete loss of crown structure. Often it was so extensive it resulted in pulpal exposure, necrosis of the pulp and subsequent apical infection.

The primary cause of this tooth wear was the chewing throughout life of a coarse fibrous diet made even more abrasive by the introduction of inorganic particles, particularly into bread, a staple food. Analysis of ancient Egyptian bread has determined that these particles were predominately quartz (sand) with the presence of some feldspar, mica, hornblende and other rock fragments (Leek, 1972, p.131). Wind-blown sand would have been the major contaminant, but particles would also have come from soil in which the grain was grown, mud-brick silos used to store the grain, stones used to grind the grain and during the process of baking (Forshaw, 2009, p. 421; Gamza and Irish, 2010).

Dental caries is a disease process that can also result in necrosis of the dental pulp and apical infection. Caries is caused by the breakdown of fermentable carbohydrates (the mono- and disaccharides such as glucose, fructose and sucrose) in the diet by bacteria found in plaque deposited on teeth. This results in the production of acid, chiefly lactic acid, which demineralises tooth enamel producing cavitation. Bacteria are then free to penetrate the enamel, progress through the dentine via the dentinal tubules and then invade the pulp chamber. Although fairly prevalent in modern societies caries was infrequently seen in ancient Nubia due to the lack of these fermentable carbohydrates in the Nubian diet. Additionally, the fibrous nature of the food tended to inhibit the retention of plaque on the tooth surface (Rateutschak-Pluss and Guggenheim, 1982, pp. 239-244). Tooth wear was also a factor since the heavy occlusal wear would have eliminated pits and fissures on the tooth surface, whilst the interproximal wear would have caused flattened tooth contacts, therefore producing a more difficult environment for plaque and caries to proliferate in (Harris et al.,1998, pp.61-62).

Trauma to a tooth, such as a dislocation, can rupture the apical blood vessels supplying the specialised connective tissue within the pulp chamber. This subsequent lack of blood supply to the pulpal tissue causes necrosis and again can result in an apical infection.

All these conditions introduce oral bacteria into the dental pulp which would then colonise the root canal chamber with what is a diverse mix of anaerobic bacteria. The bacteria and their toxic products subsequently enter the periapical tissues and induce an acute or chronic inflammatory result. The level of inflammatory response depends on the balance between the immunity of the host and the virulence of the infection.

Low-grade persistent infection produces a chronic inflammatory reaction, initially resulting in a granuloma, which may become a chronic abscess or develop into a periapical cyst. The cysts have smooth cavity walls and despite the sometimes extensive bony resorption, these chronic lesions can be benign and possibly asymptomatic. It is not uncommon to observe multiple periapical cavities in a skull, and although the infection had obviously persisted during life, the individual was not necessarily acutely ill (Dias and Tayles, 1997, pp.553-554). However, the chronically infected lesions would have been continuously draining pus into the oral cavity, there would have been an increase in the white blood cell count and the individual would have been debilitated to some extent.

Alternatively, if the infection is severe and involves pyogenic (pus-producing) bacteria, an acute periapical abscess will form. The main signs and symptoms during life would have been pain, swelling, erythema and suppuration, initially localised to the affected tooth. An acute abscess invades the intertrabecular spaces and vascular channels within bone and then affects the soft tissues. It does not form a bony cavity as there is not enough time for the stimulation of osteoclastic resorption. On penetrating the soft tissues the infection follows the line of least resistance until it reaches a flat surface with the abscess then rupturing and discharging pus externally (Dias and Tayles, 1997, pp.548-554; Robertson and Smith, 2009, p.155). An acute infection can establish itself secondarily in a pre-existing granuloma or cyst, and in the case of a cyst the walls of the bony cavity can then appear slightly roughened.

The acute dental abscess is frequently underestimated in terms of its morbidity and mortality. Dependent on the type, quantity and virulence of the micro-organisms the infection can give rise to osteomyelitis or could spread beyond the confines of the jaws, which left untreated, can potentially have serious complications. These include cellulitis and systemic bacteraemia, which can progress to generalised septicaemia and death. A major factor in the progression of these infections is host resistance or its impairment by systemic disease (Uluibau et al., 2005, p.75). Although teeth have been implicated in systemic illnesses since antiquity, the role of bacteria in this process was not recognised until the late 19th century (Thomas, 1908, pp.161-183).

One such major infection is Ludwig's angina which is a life-threatening cellulitis commonly originating from a second or third mandibular molar. An infection associated with these teeth can penetrate the lingual bone adjacent to their roots and then spread into the sublingual, submandibular and parapharyngeal fascial spaces. This is a characteristically aggressive condition and systemically there is fever, toxaemia and leucocytosis. Locally the rapidly spreading cellulitis causes distortion of the floor of the mouth which can result in airway compromise and asphyxiation (Kurien et al., 1997, p.263; Cawson and Odell, 2002, pp.94-95).

Cavernous sinus thrombosis is a complication of an infection arising from an upper anterior tooth, which can then track upwards via the labial or facial veins to the cavernous sinus at the base of the skull. Blood clot formation can occur, and even today with rapid intervention involving aggressive antibiotic therapy and surgical drainage there is a 50% mortality rate, with survivors often suffering the loss of an eye (Cawson and Odell, 2002, p.96).

Although there are many causes of head and neck infections, infections of dental origin have been determined as being the most common of these. Huang et al. (2004) found 50% of 185 cases of deep neck infections were dentally related; Bridgeman et al. (1995) found 53% in their 107 cases; Bross-Soriano et al. (2004) 89% in their 121 cases and Juang et al. (1989) 86% in their study of 14 cases of Ludwig's angina.

Were dental infections a major cause of fatalities in ancient Nubia? Obviously a difficult question to debate, but by looking at information from the pre-antibiotic era and examining surviving skeletal material, it may be possible to comment further. In England in the 1600s the London Bills of Mortality continually listed 'teeth' as the fifth or sixth leading cause of death (Clarke, 1999, p.11). DeWitte and Bekvalac (2010) examined skeletal remains dating from 1350-1538 AD from a medieval London cemetery and their findings indicated that oral pathologies were associated with an elevated risks of mortality. Even in the modern surgical but pre-antibiotic era, dental infections have been associated with a significant risk of death. A study by Thomas (1908) determined a death rate of between 10 and 40% of those admitted to hospital with dental infections.

Case studies

The study of the Nubian collections that the author undertook revealed many of the skulls displaying pathological bony cavities of dental origin and a number of these cases will now be considered.

CASE 1

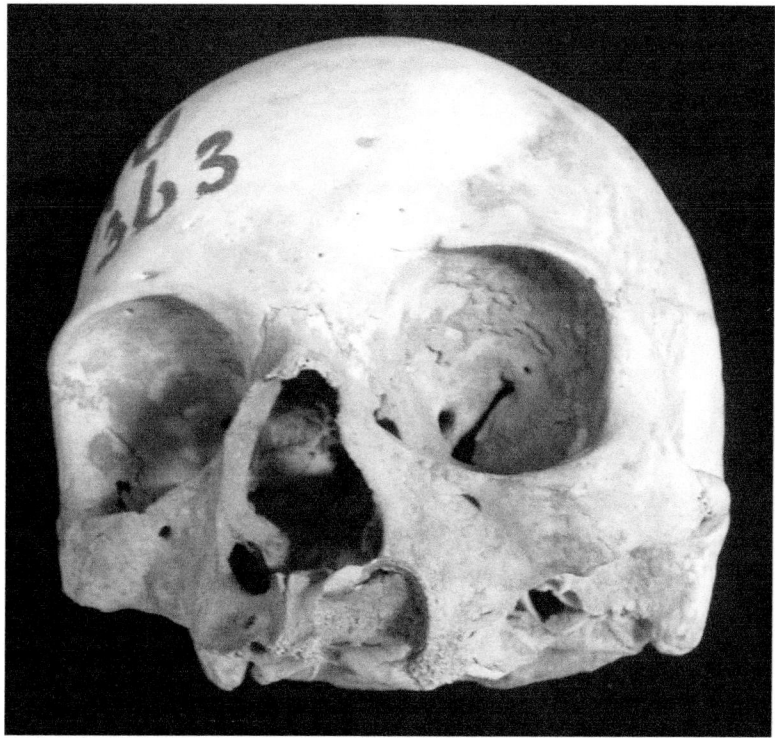

Figure 1: Skull NU363 displays a periapical cyst
in the left anterior maxillary region, measuring 18 x 15 mm.
(Courtesy of the Duckworth Laboratory, The University of Cambridge)

Case 1 displays a large periapical cyst in the maxilla of this elderly male skull (figures 1 and 2). All of the teeth have been lost antemortem and the alveolar bone is thus much reduced in size. The lesion measuring about 18 mm in diameter has a slightly roughened wall indicating evidence of chronic infection. Superiorly, the cyst has perforated the floor of the nasal cavity and so there may have been a discharge of pus into this aperture. This could have resulted in an upper respiratory tract infection and had the potential to cause a more generalised infection, particularly if the lesion had become acutely infected.

Inferiorly, the cystic lesion has resulted in considerable resorption of the alveolar and palatal bone, with the likelihood of open communication with the oral cavity by means of a fistula. There would have been a discharge of pus and the possibility of food trapping in the oral cavity.

CASE 2

This skull shows multiple pathological cavities above the roots of the maxillary anterior teeth, all of which have been lost postmortem (figure 3). The largest of these, above the

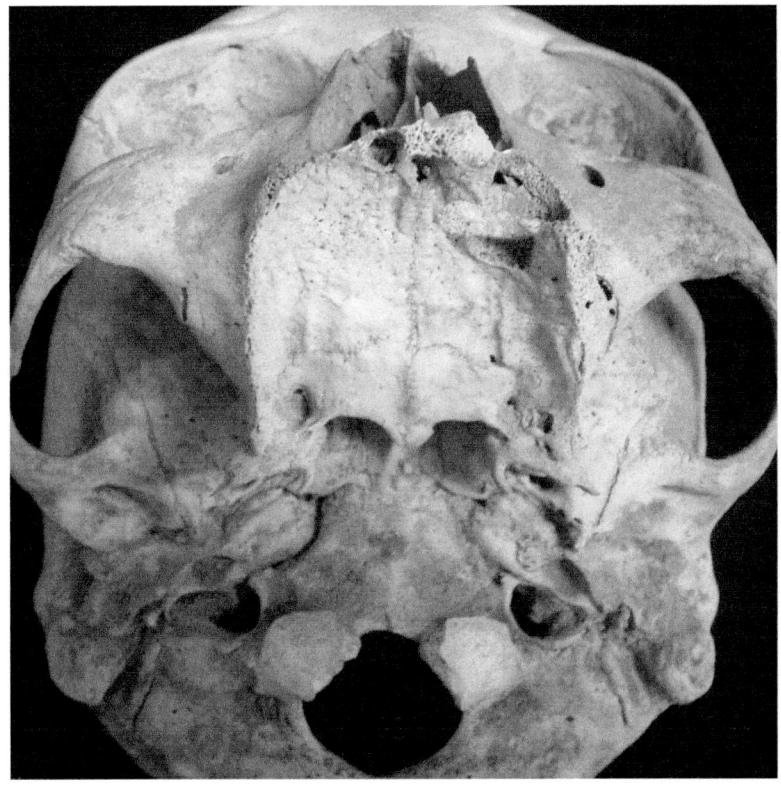

Fig. 2 Occlusal view of skull NU363
showing the extensive resorption of the alveolar and palatal bone caused by the cyst.
(Courtesy of the Duckworth Laboratory, the University of Cambridge)

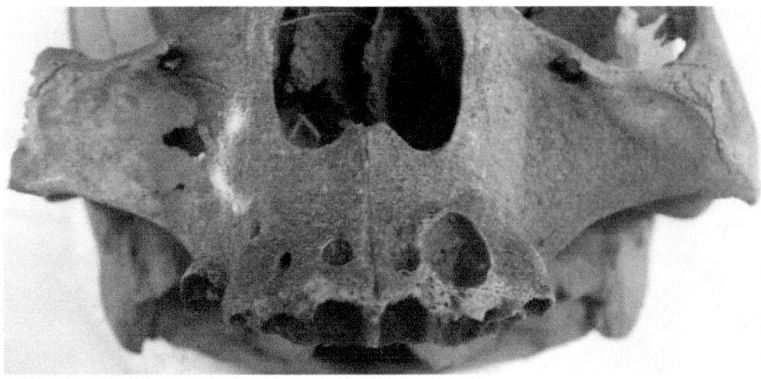

Figure 3: Multiple pathological bony cavities are evident in skull NU131.
The largest cavity associated with the maxillary left lateral incisor measures 8 x 10 mm.
(Courtesy of the Duckworth Laboratory, the University of Cambridge)

maxillary left lateral incisor, is spherical and cystic in nature. During life these lesions may well have been draining pus, and the teeth associated with them would have been sensitive to pressure during mastication. Although debilitating in nature the individual may have been healthy enough to withstand these affects. As with all such lesions they could have become secondarily affected and then entered into an acute phase of activity.

CASE 3

This skull displays a single periapical cavity above the roots of the second maxillary premolar, measuring 9 x 12 mm (figure 4). The lesion is cystic in nature with some roughness of the cavity wall and so again may have been secondarily infected. The tooth has been lost postmortem and so it is uncertain what the causative factor for the lesion would have been.

CASE 4

Skull NU441 demonstrates another example of a large cystic cavity above a maxillary molar. The cyst has smooth cavity walls and has expanded into the maxillary air sinus but there is no communication with this body. The lesion may have been relatively symptom free during life (figure 5).

CASE 5

Case 5 displays a large periapical cavity above the roots of a first maxillary molar; the cause of the infection in this case is the large distal carious lesion present in the tooth (figure 6).

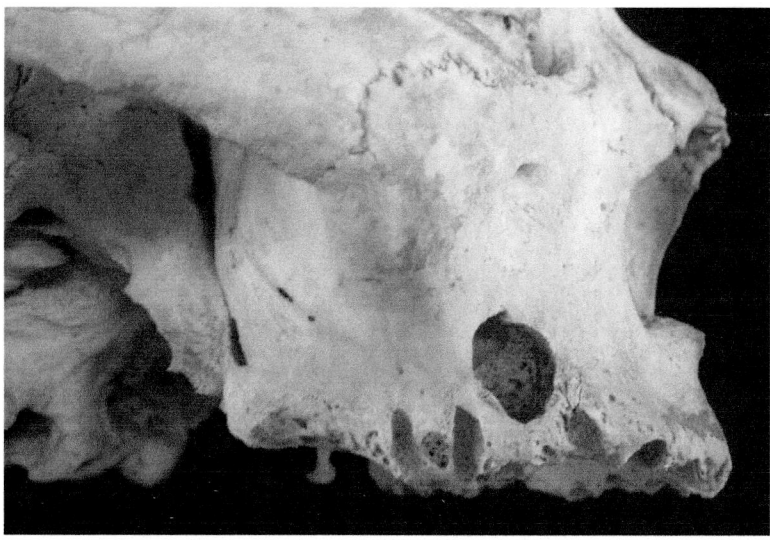

Figure 4: Skull NU322 displays a cyst associated with the second maxillary premolar.
(Courtesy of the Duckworth Laboratory, the University of Cambridge)

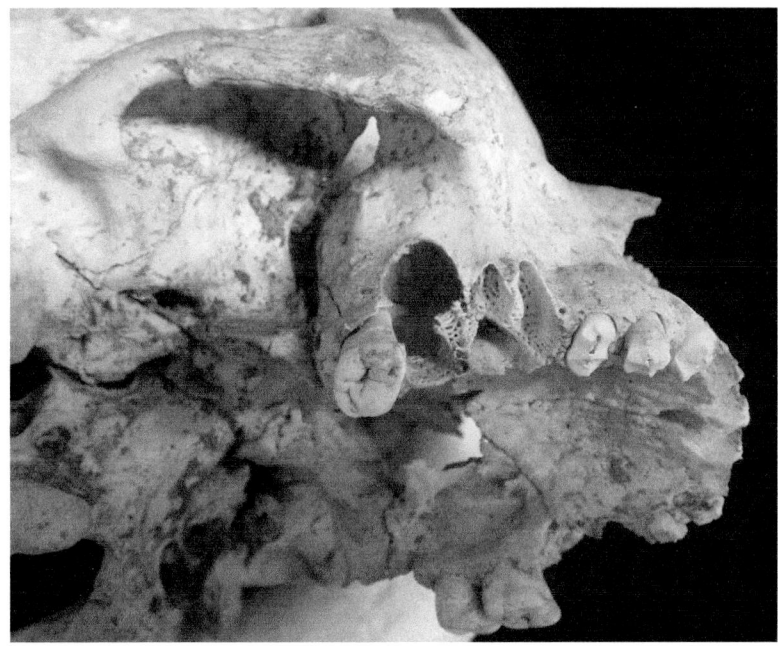

Figure 5: A cyst measuring 9 x 20 mm
associated with the maxillary second molar on skull NU441.
(Courtesy of the Duckworth Laboratory, the University of Cambridge)

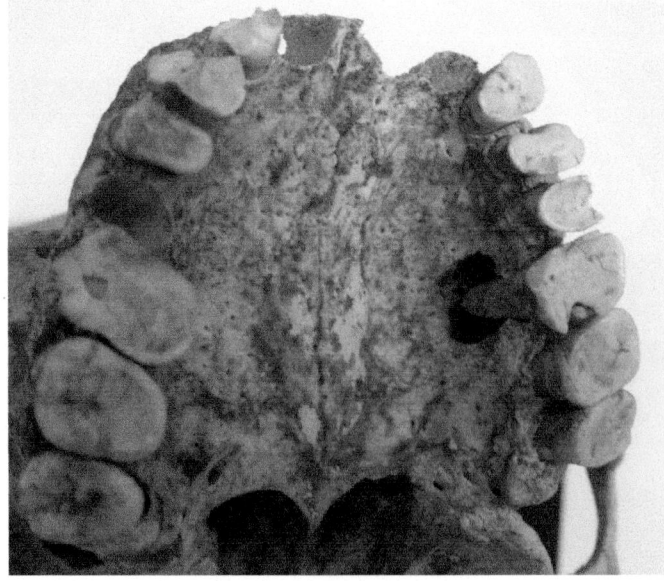

Figure 6: Periapical lesion above the roots of the first molar in skull NU616.
The lesion has perforated the inferior surface of the maxillary air sinus.
(Courtesy of the Duckworth Laboratory, the University of Cambridge)

Visual inspection indicates that the abscess has perforated the inferior floor of the maxillary sinus, with little evidence of the sinus floor remaining. As figure 7 demonstrates there is a close relationship between the roots of the posterior maxillary teeth and the floor of the sinus, with the bone constituting the sinus floor being fairly thin. As a consequence the entire sinus may have become infected, which could have resulted in serious systemic consequences. Additionally, the superior surface of the maxillary sinus forms the floor of the orbit, again with only a thin layer of bone intervening between these two anatomical cavities. Infection from the maxillary sinus could perforate this bone and an orbital abscess and cellulitis ensue, possibly resulting in blindness.

Further complications can arise from such a condition, as infection could spread to other air sinuses such as the ethmoid, frontal, and sphenoid sinuses. These sinuses are only separated from the brain by a thin layer of bone, and penetration of the bone could disseminate the infection to the fluid and tissues surrounding the brain, resulting in meningitis. Infection entering the brain tissue can cause an abscess and inflammation of the intracranial blood vessels. This can result in a thrombosis of the blood vessels inside the skull possibly leading to blindness, brain swelling, stroke, and even death (Fehrenbach and Herring, 1997; Uluibau et al., 2005).

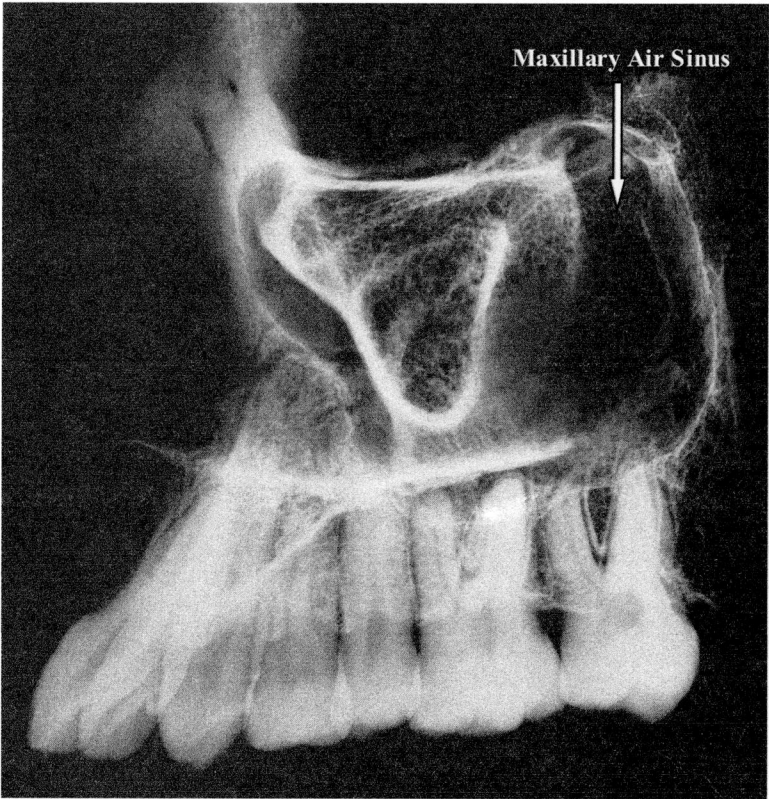

Figure 7: Radiograph showing the position of the maxillary air sinus.

CASE 6

This skull is one of a number that was excavated from the so-called 'Executioner's Trench' which is located close to the village of Shellal, near Aswan. This group of individuals, numbering about a hundred males were found to have been executed by various means such as hanging, decapitation, sword-cuts and spear-thrusts. Dated to about 250 AD, the executions were thought to have been carried out by the Romans, as there was known to have been civil unrest in the area around that time.

The skull displays a large cystic lesion associated with the right maxillary central incisor, the tooth having been lost postmortem (figures 8 and 9). Measuring about 15 mm in width the lesion has, unusually, perforated the nasal cavity and has extended inferiorly to also perforate the hard palate. Such communication would have allowed open drainage into the oral and nasal cavities. The condition would have been debilitating during life due to the discharge of pus and also uncomfortable because of the possibility of food trapping. However, the lesion was not fatal due to the known cause of death.

The remaining teeth in the dentition of this individual are in good condition with no evidence of caries, tooth wear or periodontal disease. Such observations would suggest that trauma would seem likely to be having been the cause of this condition. Some form of blow to the tooth, possibly many years previously, would have caused severing of the apical vessels supplying the connective tissue of the pulp; this would have resulted in pulpal necrosis, periapical infection and eventually a large cystic lesion.

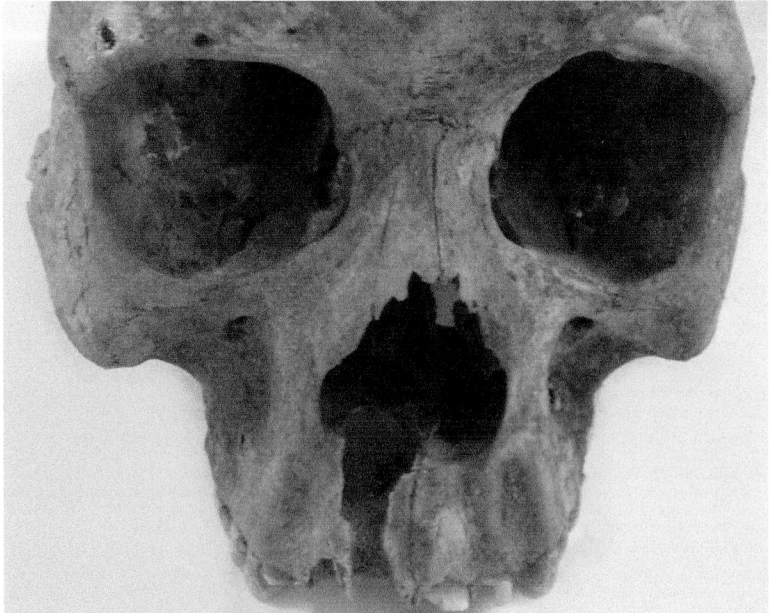

Figure 8: Skull NU737 is from one of the individuals excavated from the 'Executioner's Trench'. Displaying a cystic lesion associated with the maxillary incisor which communicates with both the oral and nasal cavities.
(Courtesy of the Duckworth Laboratory, the University of Cambridge)

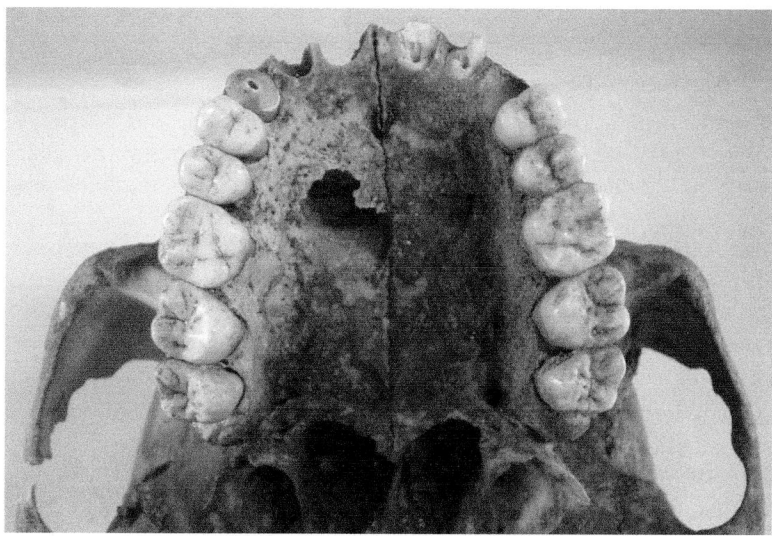

Figure 9: Palatal view of the cyst in skull NU737
(Courtesy of the Duckworth Laboratory, the University of Cambridge)

Summary and conclusion

Historical information suggests that dental infections were a significant cause of deaths in past times, whilst more modern research indicates that infections of dental origin are the most common type of head and neck infections. Even today with modern surgical treatment and aggressive antibiotic therapy, dental infections can still be fatal. Such evidence would suggest that dental disease in ancient Nubia would have been a serious source of infection and have resulted in the deaths of many individuals.

This paper examined and discussed the aetiology and possible consequences of various periapical lesions of dental origin. A number of examples of such lesions observed in the skeletal material excavated by the first Archaeological Survey of Nubia were examined and discussed. Although it is impossible to identify any of these cases as a direct cause of death, it is likely that secondary infection of such lesions would have had serious systemic consequences and may indeed have proved fatal.

Acknowledgements

I would like to thank the following for their kind permission and assistance in studying the collections in their care: Professor Norman MacLeod and Dr. Robert Kruszynski of The Natural History Museum, London; Professor Rosalie David of the KNH Centre, The University of Manchester, and Dr. Marta Lahr and Maggi Bellatti of the Leverhulme Centre for Human Evolutionary Studies, the University of Cambridge.

Also thanks to the sponsors of the project. This work was supported by The Wellcome Trust [WT090575MA].

References

Bridgeman, A., Wiesenfeld, D., Hellyar, A. and Sheldon, W., 1995. Major maxillofacial infections: An evaluation of 107 cases. *Australian Dental Journal*, 40, pp.281-288.

Bross-Soriano, D., Arrieta-Gomez, J. R., Prado-Calleros, H., Schimelmitz-Idi, J. and Jorba-Basave, S., 2004. Management of Ludwig's angina with small neck incisions: 18 years experience. *Otolaryngology-Head and Neck Surgery*, 130, pp.712-717.

Brothwell, D., 1963. *Digging up Bones*. Ithaca, New York: Cornell University Press.

Buikstra, J. E. and Ubelaker, D. H., 1994. *Standards for Data Collection from Human Skeletal Remains*. Fayetteville, Arkansas: Arkansas Archaeological Survey.

Cawson, R. A. and Odell, E. W., 2002. *Oral Pathology and Oral Medicine*. Edinburgh and London: Churchill Livingstone.

Clarke, H. J., 1999. Toothaches and death. *Journal of the History of Dentistry*, 47, pp.11-13.

DeNiro, M. J., 1987. Stable isotopy and archaeology. *American Scientist*, 75, pp.182-191.

DeWitte, S. N. and Bekvalac, J., 2010. Oral health and frailty in the medieval cemetery of St. Mary Graces. *American Journal of Physical Anthropology*, 142, pp.341-354.

Dias, G. T. and Tayles, N., 1997. 'Abscess Cavity' - a misnomer. *International Journal of Osteoarchaeology*, 7, pp.548-554.

Dupras, T. L., Schwarcz, H. P. and Fairgrieve, S. I., 2001. Infant feeding and weaning practices in Roman Egypt. *American Journal of Physical Anthropology*, 115, pp.204-212.

Fehrenbach, M. J. and Herring, S. W., 1997. Spread of dental infection. *Practical Hygiene*, Sept/Oct, pp.13-19.

Forshaw, R. J., 2009. Dental health in ancient Egypt. *British Dental Journal*, 206, pp.421-428.

Gamza, T. I. and Irish, J., 2010. A comparison of archaeological and dental evidence to determine diet at a predynastic Egyptian site. *International Journal of Osteoarchaeology*, 22(4), pp.398-408.

Harris, J. E., Ponitz, P. V. and Ingalls, B. K., 1998. Dental health in ancient Egypt. In A. Cockburn, E. Cockburn and T. A. Reyman, eds. *Mummies, Disease & Ancient Cultures*. Cambridge: Cambridge University Press. pp.59-68.

Hillson, S. W., 1979. Diet and dental disease. *World Archaeology*, 11, pp.147-162.

Huang, T. T., Liu, T. C., Chen, P. R., Tseng, F. Y., Yeh, T. H. and Chen, Y. S., 2004. Deep neck infections. *Head Neck*, 26, pp.854-860.

Iacumin, P., Bocherens, H., Chaix, L. and Marioth A., 1998. Stable carbon and nitrogen isotopes as dietary indicators of ancient Nubian populations (Northern Sudan). *Journal of Archaeological Science*, 25, pp.293-301.

Juang, Y. C., Cheng, D. L., Wang, L. S., Liu, C. Y., Duh, R. W. and Chang, C. S., 1989. Ludwig's angina: an analysis of 14 cases. *Scandinavian Journal of Infectious Diseases*, 21, pp.121-125.

Kurien, M., Mathew, J., Job, N. and Zachariah, N., 1997. Ludwig's Angina. *Journal of Clinical Otolaryngology*, 22, pp.263-265.

Leek, F. F., 1967. The practice of dentistry in ancient Egypt. *Journal of Egyptian Archaeology*, 53, pp.51-58.

Leek, F. F., 1972. Teeth and bread in ancient Egypt. *Journal of Egyptian Archaeology*, 59, pp.126-132.

Miller, J., 2008. *An Appraisal of the Skulls and Dentitions of Ancient Egyptians, Highlighting the Pathology and Speculating on the Influence of Diet and Environment*. Oxford: Archaeopress.

Rateutschak-Pluss, E. M. and Guggenheim, B., 1982. Effects of a carbohydrate-free diet and sugar substitutes on dental plaque accumulation. *Journal of Clinical Periodontology*, 9, pp.239-244.

Robertson, D. and Smith, A. J., 2009. The microbiology of the acute dental abscess. *Journal of Medical Microbiology*, 58, pp.155-162.

Ruffer, M. A., 1920. A study of abnormalities and pathology of ancient Egyptian teeth. *American Journal of Physical Anthropology*, 3, pp.335-382.

Stenhouse, M. J. and Baxter, M. S., 1979. The uptake of bomb 14C in humans. In R. Berger and H. Suess, eds. *Radiocarbon Dating*. Berkley, CA: University of California Press. pp.324-341.

Thomas, T. T., 1908. Ludwig's angina: An anatomical, clinical and statistical study. *Annals of Surgery*, 47, pp.161-183.

Thompson, A. H., Richards, M. P., Shortland, A. and Zakrzewski, S. R., 2005. Isotopic paleodiet studies of ancient Egyptian fauna and humans. *Journal of Archaeological Science*, 32, pp.451-463.

Uluibau, I. C., Jaunay, T. and Goss, A. N., 2005. Severe odontogenic infections. *Australian Dental Journal Medications Supplement*, 50, pp.74-81.

A case of severe ankylosis of temporomandibular joint from New Kingdom necropolis (Saqqara, Egypt)

Ladislava Horáčková[1] and Frank Rühli[2]

[1]Division of Medical Anthropology, Department of Anatomy, Faculty of Medicine, Masaryk University (Czech Republic)
[2]Institute for Evolutionary Medicine, Institute of Anatomy, University of Zurich (Switzerland)

Abstract

One of the rare paleopathological findings discovered at the New Kingdom necropolis at Saqqara (during the 2007 season) was a case of unilateral ankylosis of the temporomandibular joint (TMJ). Human skeletal remains discovered here by the international expedition organised by the Rijksmuseum of Leiden came from excavations in the three chapels of Ptahemwia. He was named 'Royal Butler, Clean of Hands' during the reigns of the pharaohs Akhenaten and Tutankhamun (1353-1323 BC). Precise dating of the studied skeleton is nevertheless difficult because remains from the chapels found most superficially were secondary burial sites. The objects associated with these burials can mostly be dated to the late XIXth and XXth Dynasties.

The mandible of a 16-17 year old individual with a severely deformed head of the left condylar process, a sequential deformation of the mandibular fossa of the temporal bone and an asymmetry of the entire mandible is one of the most interesting finds that came from the north chapel of Ptahemwia. The right mandibular head was also stricken with a degenerative process, apparently due to a changed chewing mechanism, since the handicapped individual loaded the impaired temporomandibular joint less and thus overloaded the relatively healthy side. Ankylosis of the TMJ is most commonly associated with trauma, local or systemic infection, or systemic disease. In this article the authors discuss the anatomical particularity of the mandibular condylar process in relation to a trauma of this area and offer detailed differential diagnostics.

Introduction

The studied human skeletal remains came from the Dutch excavations of the New Kingdom Necropolis at Saqqara, Egypt. Tombs of important officers of the New Kingdom (1550–1069 BC) have been discovered at the site of the Dutch concession in the last few decades, which is located approximately 300m south of the oldest Egyptian pyramid – the pyramid of Djoser. Here were buried for example: Maya, the Overseer of the Treasury during the reign of the Pharaoh Tutankhamun, who was in charge of burials of Tutankhamun (1336–1327 BC), his successor Ay and General Horemheb, who became the last pharaoh of the 18th Dynasty; Vizier Tia and his wife Tia (who was sister of Pharaoh Ramesses II.); Meryneith, one of high officials of the Pharaoh Akhenaten (1353-1335 BC) and others. Another burial complex was discovered in the year 2007 directly to the east of the tomb of Meryneith. The tomb proved to belong to a "royal butler, clean of hands" named Ptahemwia. Previously, this person was known only from a pilaster in Bologna and a door-jamb in the Cairo Museum.

The burial complex of Ptahemwia consists of a rectangular enclosure of about 10.5 m wide and 16 m long. Most walls still stand to a height of about 2 metres. The massive entrance gateway is followed by a courtyard with a peristyle of papyriform columns (the lower part of three of these still preserved), and three chapels (the north, south and central) for the offering cult. The main scene in the east wall is that of the transport of the mummy of the deceased. The main attraction of the west half of north wall is the realistic depiction of two monkeys under the wife´s chair which are eating figs, dates, and grapes. We can only speculate why the reliefs on the north and east walls were never finished - perhaps the owner just died unexpectedly, it is also possible that he fell from favour or was transferred to Amarna. In view of the peculiar technique and style of its decoration the tomb can be dated to the reign of Akhenaten (1353-1335 BC) (Raven et al. 2007).

The aim of this article is to discuss the discovery of a temporomandibular joint (TMJ) ankylosis in the fragments of a skull and mandible of a c.16-17 year old individual buried secondarily in the north chapel. This burial can be dated to the late XIXth – XXth Dynasties.

Material and methods

Fragmented skeletons not in correct anatomical position were discovered in all three chapels of the burial complex of Ptahemwia. Precise dating of the studied skeletons is nevertheless difficult because the superficial burials found in the chapels are considered to be secondary burials. Children were mostly buried deeper in the north and south chapels. Their skeletons were nearly undamaged indicating that these were primary burials. It is possible that both of these chapels were originally intended for the burial of children. Human remains from a minimum of 24 individuals were found in the north chapel while at least 56 individuals were buried in the south chapel.

The skeletal elements found in Ptahemwia´s chapels were studied macroscopically, their dimensions and basic demographic parameters were ascertained and the age was ascertained from the dentition and the degree of fusion observed in the bones of limbs (see Martin and Saller 1957; Trotter and Glesser 1958; Howells 1964; Černý 1971; Lovejoy 1985; Ubelaker 1987; Knussmann 1988; Loth and Hennenberg 1996; Stloukal et al. 1999). The variability and pathological changes visible in the skeletons were also studied (see Brothwell and Sandison 1967, Steinbock 1976; Zimmermann and Kelly 1982; Hauser, De Stefano 1989; Aufderheide and Rodríguez-Martín 1998; Ortner 2003, Horáčková et al. 2004; Brothwell 2010).

Special attention has been paid to a partially preserved skeleton in which the mandible and temporal bone were firmly connected. Using the methods described above (mainly on the basis of dentition, obliteration of skull sutures, degree of development of the pubic symphysis relief and the preserved epiphyses of long limb bones) the age at death of this individual has been determined to be 16-17 years. The pelvis and other bone morphologies presented predominantly male characteristics. The discovery of TMJ pathology was the reason why the bones of the splanchnocranium of this individual have been studied very carefully.

Description of the temporomandibular joint (TMJ)

The TMJ is a complex joint with articular facets on the heads of the mandible and mandibular fossa and the articular tubercle of the temporal bone. The TMJ is divided into two compartments by the articular disc. The disc is attached to the margins of the joint capsule, and to the pterygoid lateral muscle. The longitudinal axis of both mandibular heads converge to the median line at an angle of 150-160°. Functionally, the TMJ represents a combination of two joints: an articulation between the articular disc and the head of the mandible and the articulation between the articular disc and the mandibular fossa. Active opening of the mouth always involves a rotary movement at the lower joint and a sliding movement anteriorly at the upper joint. The TMJ is a "double-side" joint, which means that a movement on one side is not possible without a movement of joint on the other side (Čihák, 1987; Platzer, 1992).

Particularity of the temporomandibular joint

The mandible plays a special role in the development of the facial part of the skull and the mechanism of mandible fracture repair is also unique in the body. Meckel's cartilage is the basis of the mandible in the human embryo – the cartilage being a column of dense embryonic cartilage stretching through the mandibular protuberance of the face from the chin region to the future middle ear cavity where it passes to the basis of the malleus without any sharp transition. A bone plate arises by enlarging the bone base externally of Meckel's cartilage, which proliferates under Meckel's cartilage. It forks externally of the cartilage, so that the desmogenous ossification makes a sort of a bone trough opening upwards, which looks like a letter Y in section. A relatively strong nervous branch and a strong artery with a vein (future inferior alveolar vessels and nerve) lie between the short arms of this formation. Blood vessels penetrate into Meckel's cartilage in the area of the future anterior teeth in parallel with this process. Cells on their surface destroy the substance of Meckel's cartilage in the way typical for enchondral ossification. Meckel's cartilage resorbs fast and is replaced with bone tissue, which blends with the bone of the mandible in a desmogenous way. All these processes are located in the area of the future body of the mandible.

The origin of the mandibular ramus is different. A column of cartilage appears in the area of the future mandibular ramus somewhat later than Meckel's cartilage. This column is surrounded by bone tissue from the neighbouring connective tissue and undergoes enchondral ossification itself that proceeds from beneath – from the future mandibular angle – upwards to the basis of the mandibular joint. The ossification proceeds at the end of the foetal and at the beginning of postnatal periods of life to the area of future mandibular neck only. It stops there and cartilage cells re-align themselves in the space so that the cartilaginous mandibular head is fixed to the newly made bone by strips of cartilage and ligament resembling mast anchoring cables. The boundary between these two tissues takes over the function of growth cartilage, from which the mandibular ramus grows to height. This boundary is less strong mechanically. It may happen in injuries where a child falls on their chin and the mandible neck fractures in the mechanically least resistant place. The mandibular neck is subluxed and rotated medially by the pull of muscles. It is not in contact with the mandibular fossa and it resorbs rather fast. A new head can originate from the growth cartilage in the mandibular neck, resulting in a new temporomandibular head. This mechanism of fracture repair is unique in the body. The resorbtion of the broken mandibular head and the creation of the substitute head takes some time (several months), in which the mandibular ramus growth slows down or stops temporarily or permanently (depending on the age of the injured child). The asymmetry of the mandibular apparatus and the face originates from that (Mrázková and Doskočil 1994).

Results and discussion

Description of the lesion

Both heads of the mandible and mandibular fossa are changed by a pathological process (figure 1). The left head of the mandible has complete loss of the joint surface from erosion and abrasion and is flattened irregularly. Its articular surface has grown together tightly with the mandibular fossa and the articular tubercle on the lateral side. There is an adhesion also between the extended root of the zygomatic process of the temporal bone and the altered mandible head on its ventrolateral side. However, the adhesion field was disturbed by fragments of the skull bones during postmortem handling (figure 2). The extended area at the root of the zygomatic process of the temporal bone is 21 mm long in the anteroposterior direction and approximately 8 mm high. There is a recess in the place, where it continues in the supramastoid crest, into which one of lobed protuberances of the affected head locks. The bottom of the mandibular fossa and the articular tubercle is porous and rough (figure 3). The tympanic plate and the entrance to the external acoustic porus are thicker and they are covered by newly created bone tissue (a possible sign of a periostitis). A small perforation is apparent approximately in the disc centre.

The affected head inclines dorsally, its transversal dimension amounts to 23 mm and the anteroposterior one to 22 mm (figure 4). A protuberance 17 mm long and 8 mm wide, covered by compact bone, projects medioventrally. The dorsal end of the head extends beyond the dorsal end of the mandibular ramus by approximately 8 mm. Apices of both coronoid processes are 91 mm apart. The lingula of the mandible is strengthened into a protuberance of pyramidal shape on the inside of the left mandibular ramus. A pointed 18 mm long protuberance projects in the cranial direction from the pterygoid tuberosity. This protuberance can be regarded as a myositis ossificans, where part of the pterygoid medial muscle fibres were overloaded due to impaired joint mechanics unusually for a long time. The mylohyoid groove lies in a nearly horizontal position, only 8 mm above the upper edge of the mandibular body. The mylohyoid line makes a protruding elevation. The mental foramen is situated between the second premolar (P2) and the first molar (M1). The mandibular angle creates a protuberance, which is elongated in the caudal direction noticeably.

The right condylar process is also affected by the pathological process; the lesion is of a much smaller extent, however (figure 5). The mandibular head (maximum length 16 mm, width 15 mm) is flattened and semicircular in the medial half; the surface of the

Figure 1: Fragments of the skull of a 16-17 years old individual with completely destroyed articular surfaces of temporomandibular joints by a pathological process. (Tomb of Ptahemwia, New Kingdom Necropolis, Saqqara, Egypt). Photo by Anneke J. de Kemp.

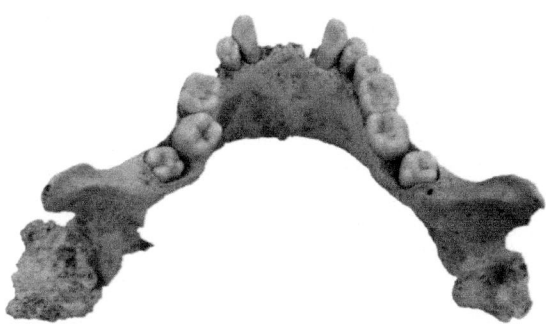

Figure 2: Mandible of a 16-17 years old individual from the north chapel of tomb of Ptahemwia. Both heads of the mandible are changed by a pathological process. (Tomb of Ptahemwia, New Kingdom Necropolis, Saqqara, Egypt). Photo by Ladislava Horáčková.

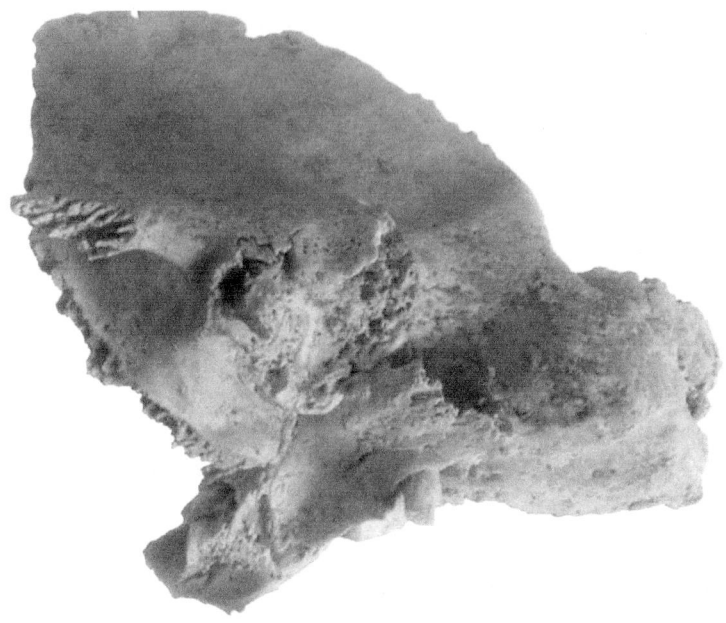

Figure 3: Porous and rough bottom of the left mandibular fossa and articular process with numerous small perforations. The tympanic plate and entrance to the external acoustic porus are thicker and they are covered by newly formatted bone tissue. (Tomb of Ptahemwia, New Kingdom Necropolis, Saqqara, Egypt). Photo by Anneke J. de Kemp.

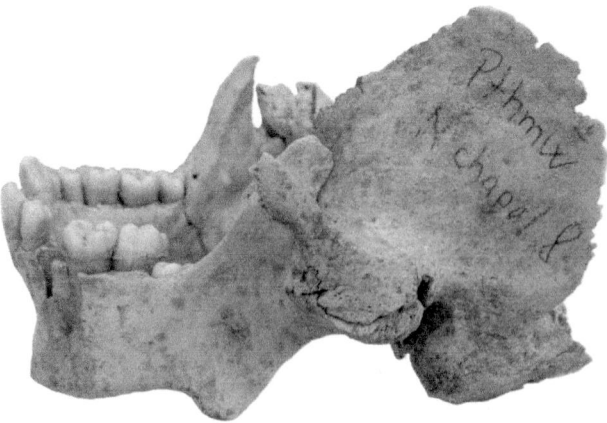

Figure 4: View to the left temporomandibular joint with ankylosis. (Tomb of Ptahemwia, New Kingdom Necropolis, Saqqara, Egypt). Photo by Ladislava Horáčková.

lateral half of the joint is furrowed irregularly and covered by compact bone. The biggest protuberance newly formed on the condylar process is elongated medially. The head and the mandibular fossa grew together completely in the dorsolateral quarter of this part. This adhesion was disturbed postmortally as in case of the left TMJ. The mylohyoid line is pronounced as well as the submandibular fossa, which has a maximum medial concavity.

The majority of the bottom of the right mandibular fossa is rough on the lateral side and the articular tubercle is almost completely destroyed. The surface is covered by compact bone tissue. The tympanic plate of the temporal bone has signs of slight periostitis. There is a tiny opening apparent approximately in the centre.

The mental spine does not lie in the median plane; it has a slight lateral inclination. The asymmetry can best be seen at the medial parts of the sublingual foveae on both sides. The left chin area is lower (mental protuberance) from the outside. The left mandibular angle is about 140°, the right one 135°. A slight eversion is apparent at both angles.

The alveolar process is higher in the medial direction so that the teeth make a high arch in the middle part of the mandibular body, which lowers in the direction of the molars (the mandibular body height is 28 mm in the second molar (M_2) left point, 34 mm in symphysis, 27 mm in the second molar (M_2) right point, 22 mm at the third molar (M_3) right as well as left).

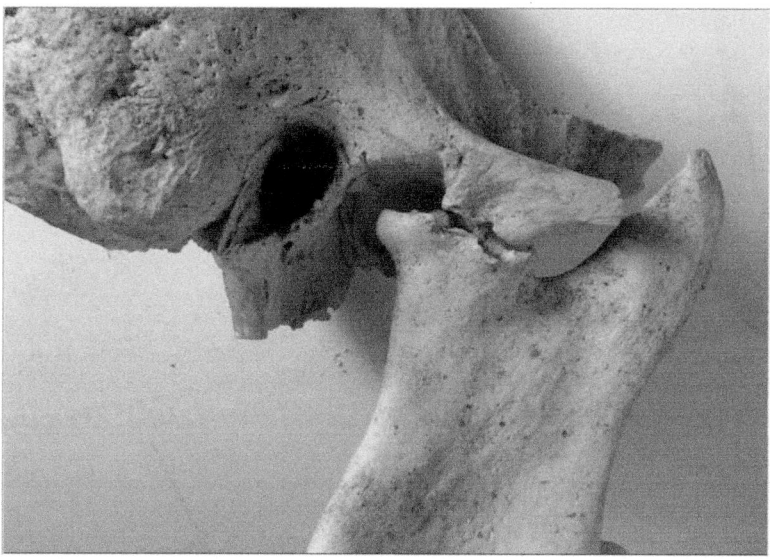

Figure 5: The affected right temporomandibular joint.
(Tomb of Ptahemwia, New Kingdom Necropolis, Saqqara, Egypt).
Photo by Anneke J. de Kemp.

The canines protrude noticeably above the first premolar (P_1) on both sides and the right canine rotates slightly in a lateral direction. The teeth are cramped in the dental arch, especially the right canine and the first premolar, where the canine erupted lingually from the lateral incisor (I_2) and the first premolar (P_1). The third molars (M_3) are just erupting on both sides, their crowns protruding approximately 2mm above the edge of the alveolus. Abrasion is more distinctive on the less affected right side of the dental arch which was used somewhat longer than the left side probably due to mechanical conditions and soreness. Abrasion began at the first premolar (P_1) on the right side. It affected the vestibular cusp mainly; both cusps - vestibular as well as lingual – are slightly abraded at the second premolar (P_2). The tips of all cusps are abraded at the first molar (M_1) of the right side, so that isolated dentine areas appeared. Slight enamel abrasion is present at the vestibular cusps only at the right side the second molar (M_2). The developing third molar (M_3) rather swerves from the line of occlusion in a lateral direction. The enamel is abraded at the vestibular cusps of the left first molar (M_1). The left second molar (M_2) is rotated in a medial direction so that the vestibular area is oriented cranially and the occlusal tooth surface is almost parallel with the median plane. There is almost no enamel abrasion. Thus, the abrasion at the right side is adequate to the determined age, while it is minimal at the left side.

Caries are noticeable between both central incisors (I_1 right and I_1 left) at the mesial side. They extend from the middle of the crown towards the roots and reach to the pulp cavities of both teeth.

There is a cavity at the root of the right incisor, the bottom of which is smooth with several tiny perforations. The cavity is open outside by an oval opening (11 x 8 mm) with smoothed rims. Obviously, this was a consequence of chronic suppurative inflammation (dentoalveolar cyst) of the dental alveolus of this incisor. A fine periosteal deposit of newly formatted bone tissue is noticeable in the vicinity of the lesion. Traces of gingivitis in the form of tiny pits are evident around the edges of the alveolar processes of the incisors on both sides. A thick deposit of dental plaque is apparent on the incisors and canines from the lingual side above all and from the vestibular, as well as the lingual side on the remaining teeth.

Conclusion/diagnosis: Temporomandibular joint ankylosis, asymmetry of mandible, dentoalveolar cyst.

Etiology of temporomandibular ankylosis

Temporomandibular joint ankylosis is a pathological process caused by damage of the articular facets (head of mandible and mandibular fossa with articular process of temporal bone) and their fusion. It can be regarded as the terminal phase of some diseases.

Hypomobility of the mandibular joint is a basic clinical symptom of TMJ ankylosis. Opening the mouth is possible, usually within a range of 5-7 mm. The articular head makes the indicated rotational movement only, and no translation is possible. Neither protrusion (shifting the jaw in the anterior direction) nor laterotrusion (moving the jaw to the side) are possible. These limitations to jaw movement are connected to complications with food intake, insufficient oral hygiene, and impaired diction and, eventually, breathing.

The damage to the temporomandibular joint may be due to several causes. They are most frequently:

1. Macrotrauma – contusion and fracture (with bleeding to intraarticular space). Joint ankylosis is also believed to be caused by repeated episodes of bleeding, as found in patients diagnosed e.g. with haemophilia.
2. Avascular necrosis of the mandibular condyle, metabolic diseases or congenital conditions.
3. Microtrauma (development of degenerative osteoarthritic changes due to the wear and tear of the TMJ).
4. Tumours.
5. Systemic diseases (e.g. rheumatoid arthritis – the TMJ is affected in about 25% of cases, but juvenile chronic arthritis, the juvenile form of rheumatoid arthritis, avoids the TMJ).
6. Infection - transfer of inflammation from the tympanic cavity (for example otitis media) or viral infection from the parotid gland (for example a complication of mumps).

When ankylosis of the TMJ takes place in subjects during their developmental stage, it results in an alteration of the entire maxillofacial complex. The bone growth alterations and face asymmetry occur at the affected side. The microgenia of the lower jaw with a bird-like profile of the face – "bird face" – typically develops in cases of double-sided ankylosis.

Figure 6 shows the X- ray of the affected bones of the individual from the Ptahemwia´s tomb.

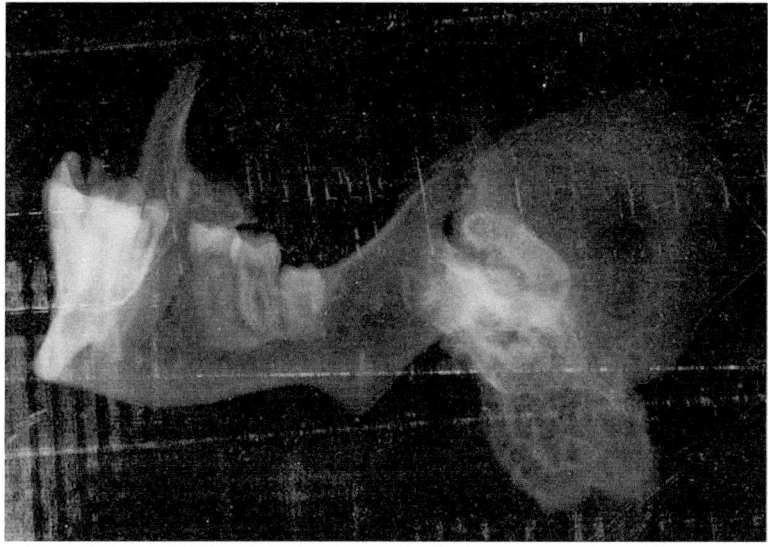

Figure 6: X-ray photograph of affected temporomandibular joint
- fracture traces have not been noticed. (Tomb of Ptahemwia,
New Kingdom Necropolis, Saqqara, Egypt).
X-ray by Salima Ikram, American University Cairo.

Fracture traces have not been noticed on the X-ray photograph, leading to a fracture being ruled out as the cause of the ankylosis (though to be absolute precise CT is required). X-rays have been done thanks to Professor Salima Ikram, American University Cairo, using a portable X-ray machine provided by the Institute for Bioarchaeology. Likewise, we have also excluded avascular necrosis, where the death of bone tissue results from a lack of blood supply, which can lead to tiny breaks in the bone and its eventual collapse. The diagnosis of developmental and endocrine disturbances from only fragments of the skeleton is not possible.

Osteoarthritis (OA) is a disease associated with the ageing process (Brothwell and Sandison 1967). Since the TMJ is a paired joint that cannot function alone, osteoarthritis, if present, (usually in old people), is often symmetrical. Lesions are more severe in the temporal surface than in the mandibular condyle (Mann and Hunt, 2004). In contrast, the studied individual from Ptahemwias's tomb is young and the lesions affected mainly condylar process of the mandible.

No signs of any tumour processes have been found.

Infection in a joint may also lead to ankylosis, depending on damage to connective tissues such as the bone, cartilage, fluid, blood vessels and nerves found within a joint.

Systemic and local infections (30-50%), such as osteomyelitis, mumps and eventual transmission of the middle ear inflammatory process into the joint area (however rare in the current 'antibiotic' era), are further causes of ankylosis formation. A continued inflammation of connective joint tissues destroys synovial membrane, cartilage and chondral bone structures. The resulting damage brings stiffness and reduces mobility in the joint.

Middle ear inflammation is one of the most common diseases in childhood. This is attested to by present-day medical statistics which demonstrate that visits to the doctor because of middle ear problems are most frequent between the 4th and 5th years of life and represent almost 40% of all diagnoses made during this period. Even today, 51,000 children younger than five years die of otitis media in developing countries according to information from the World Health Organisation (WHO) (2004).

Ear diseases are known from ancient Egypt. Their diagnosis and treatment are mentioned for example in the Berlin papyrus as well as in the Ebers papyrus (Halioua, 2002).

Based on these considerations we think that osteomyelitis is the most likely origin of this process as the first (and chronic) infection and severe ankylosis as the serious complication and terminal stage.

Conclusion

Temporomandibular joint ankylosis in children is a challenging problem even today. Surgical correction is technically difficult and the incidence of recurrence after treatment is high.

In literature concerned with paleopathology, references to the TMJ ankylosis are rare, so the described temporomandibular joint ankylosis is a rare finding in skeletal remains from ancient Egypt. It is clear from the degree of alteration to the bone that it was a long-standing chronic process that demanded care of the patient.

Even though this is just casuistry, it is possible to judge from this isolated case that the ancient Egyptian society must have taken care of even handicapped individuals, as the individuals thus affected would not be able to survive for a long time without the help of others.

Acknowledgements

FR has been supported by the Mäxi Foundation, Zurich, Switzerland.

References

Aufderheide, A.C. and Rodríguez-Martín, C., 1998. *The Cambridge encyclopedia of human Paleopathology*. Cambridge: Cambridge University Press.

Brothwell, D., 2010. On Problems of Differential Diagnosis in Palaeopathology, as Illustrated by a Case from Prehistoric Indiana. *International Journal of Osteoarchaeology*, 20, pp.621-622.

Brothwell, D. R. and Sandison, A. T. eds., 1967. *Diseases in Antiquity*. Springfield, IL: C.C. Thomas.

Čihák, R., 1987. *Anatomie 1*. Avicenum: Zdravotnické nakladatelství Praha. p.114.

Černý, M., 1971. Určování pohlaví podle postkraniálního skeletu. In: *Symposium o určování stáří a pohlaví jedince na základě studia kostry*. Praha: Národní muzeum. pp. 46-62.

Halioua, B., 2002. *Medicína v době faraonů. (La médecine au temps des pharaons)*. Nakladatelství Brána. pp.74-75.

Hauser, G. and De Stefano, G.F., 1989. *Epigenetic variants of the human skull*. Stuttgart: Lubrecht and Cramer Ltd.

Horáčková, L., Strouhal, E. and Vargová, L. 2004: *Základy paleopatologie*. Panoráma biologické a sociokulturní antropologie. Brno: Nadace Universitas Masarykiana, Edice Scientia.

Howells, W.W., 1964. Détermination du sexe du bassin par fonction discriminante. *Bulletins et Mémoires de la Société d´Anthropologie*, 7, pp.95-105.

Knussmann, R., 1988: *Anthropologie. Handbuch der vergleichenden Biologie des Menschen. Band I*. Wesen und methoden der Antropologie. Stuttgart: Gustav Fischer Verlag.

Loth, S.R. and Hennenberg, M., 1996. Mandibular Ramus Flexure: A New Morphologic Indicator of Sexual Dimorphism in the Human Skeleton. *American Journal of Physical Anthropology*, 99(3), pp.473-485.

Lovejoy, C.O., 1985. Dental Wear in the Libben Population: Its Pattern and Role in the Determination of Adult Skeletal Age at Death. *American Journal of Physical Anthropology*, 68(1), pp.47-56.

Mann, R.W. and Hunt, D.R., 2004. *Photographic Regional Atlas of Bone Diseases.* Springfield, II: C.C. Thomas. pp.60-62.

Martin, R. and Saller, K., 1957. *Lehrbuch der Anthropologie in systematischer Darstellung. Band I.* Stuttgart: Gustav Fischer Verlag.

Mrázková, O. and Doskočil, M., 1994. *Klinická anatomie pro stomatology.* Praha: Nakladatelství a vydavatelství Alberta. pp.49-54.

Ortner, D.J., 2003. *Identification of pathological conditions in human skeletal remains.* London: Academic Press.

Platzer, W., 1992. *Locomotor System. Color Atlas of Human Anatomy, Vol. 1.* New York: Georg Tieme Verlag.

Raven, M.J., van Walsem, R., Aston, B.G., Horáčková, L. and Warner, N., 2007. Preliminary report on the Leiden excavations at Saqqara, season 2007: The tomb of Ptahemvia. *Ex Oriente Lux,* 40, pp.19-39.

Steinbock, R.T., 1976. *Paleopathological diagnosis and interpretation.* Springfield II: C. C. Thomas.

Stloukal, M., Dobisíková, M., Kuželka, V., Stránská, P., Velemínský, P., Vyhnánek, L. and Zvara, K., 1999. *Antropologie. Příručka pro studium kostry.* Praha: Národní museum. pp.387-389.

Trotter, M., and Glesser, G.C., 1958. A re-evaluation of estimation of stature based on measurements of stature taken during life and of long bones after death, *American Journal of Physical Anthropology,* 16, pp.79-123.

Ubelaker, D.H., 1987. Estimating Age at Death from Immature Human Skeleton: An Overview. *Journal of Forensic Sciences,* 32(5), pp.1254-1263.

WHO, 2004. *Chronic suppurative otitis media: Burden of illness and management options.* [online] World Health Organisation. Available at:
http://www.who.int/pbd/deafness/activities/hearing_care/otitis_media.pdf

Zimmermann, M.R. and Kelley, M.A., 1982. *Atlas of Human Paleopathology.* New York: Praeger.

Occlusal macrowear, antemortem tooth loss, and temporomandibular joint arthritis at Predynastic Naqada

Nancy C. Lovell

Department of Anthropology, University of Alberta, Edmonton, Canada

Abstract

This paper is based on the results of an examination of crania and mandibles from three cemeteries at Predynastic Naqada, which were excavated by Petrie in 1895. These remains are curated as part of the Duckworth Collection at the University of Cambridge. Patterns of occlusal macrowear, antemortem tooth loss, and lesions of the temporomandibular joint (TMJ) are described, and are discussed in the contexts of diet and the biomechanics of mastication. The incomplete nature of most of the dentitions restricted the assessment of the pathological conditions, but no statistically significant differences were observed in the prevalence of TMJ arthritis between males and females, nor between elite and non-elite cemetery samples. Furthermore, antemortem tooth loss and occlusal wear were not associated with TMJ lesions.

Introduction

A workshop reviewing a century of palaeopathological research in Egypt and Nubia would not be complete without acknowledging the value of the skeletal collections acquired through the fieldwork undertaken by Sir William Mathew Flinders Petrie (1853-1942). Although it's true that Petrie was not concerned with the analysis of human remains himself, he supplied his colleagues at the University of Cambridge with skeletal remains from his excavations to aid their research in craniomorphological variation. This served to establish a number of rare collections dating from the Predynastic to the later Pharaonic eras.

One of these collections stems from Petrie's excavations at Naqada, which he undertook in 1894-1895, assisted by J. E. Quibell (Petrie and Quibell, 1896). Petrie shipped to Cambridge a portion of the material he excavated at Naqada for study by members of Karl Pearson's Biometrics School (e.g., Warren, 1897; Fawcett and Lee, 1902), but the continued curation of the remains at the University of Cambridge has made possible more recent analyses of biological affinities, diet, and health (e.g., Keita, 1990; Bartell, 1994; Johnson and Lovell, 1994; Prowse and Lovell, 1996; Keita and Boyce, 2001; Greene, 2006; Miller, 2008). Indeed, curated skeletal remains have provided the source of data for many of the foundational works in Egyptian and Nubian palaeopathology (e.g., Brothwell, 1963), and reviewed most recently by Forshaw (2009).

In this paper I describe patterns of occlusal macrowear, antemortem tooth loss (AMTL), and lesions of the temporomandibular joint (TMJ) among the Predynastic Egyptians from Naqada, and discuss these patterns in the contexts of diet and the biomechanics of mastication. Rather than cover all aspects of dental disease, I have restricted the scope of this paper so that the focus is on lesions of the TMJ and the dental conditions that may be implicated in the cause of those lesions.

Materials and methods

Petrie identified three cemeteries at Naqada and the pottery from these cemeteries provided him with some of the most important data upon which he based a new dating method and a chronology for Predynastic Egypt that bears the site's name. Skeletal remains from these cemeteries form part of the Duckworth Collection at the University of Cambridge, which now is housed at the Leverhulme Centre for Human Evolutionary Studies.

In this study, occlusal macrowear, AMTL, and lesions of the temporomandibular joint were recorded from remains representing a total of 123 adults (table 1) from the Great New Race Cemetery (usually known as Cemetery G, but also referred to as Cemetery N), Cemetery B (named after nearby Kom Bellal), and Cemetery T (named for its proximity to two tumuli). As can be seen in table 1, fewer than half of this total had complete skulls preserved. Furthermore, many of the preserved remains were incomplete in one form or another, either missing portions of the cranium or mandible due to damage unassociated with pathological conditions, or missing teeth that had been lost postmortem.

Cemetery	# Crania with associated mandibles	# Crania without mandibles	# Mandibles without crania	Total
G	29	19	16	64
B	9	17	10	36
T	5	11	7	23
Total	43	47	33	123

Table 1: The Naqada skeletal sample of adults available for examination of the teeth and alveolar bone and the components of the temporomandibular joint.

The postmortem loss of teeth can be attributed to problems of preservation (which is not uncommon when dealing with archaeological human remains in many parts of the world, particularly those remains that have not been protected in burial shafts or containers) and to loss of teeth during excavation and subsequent handling. In this sample, the single rooted incisors and canines were the teeth most commonly lost postmortem.

Estimation of the age at death for each individual followed standard bioarchaeological protocols. Unfortunately, age estimation methods for adults are highly imprecise, a situation that is made more difficult when only the crania are available for examination. In this study, ectocranial suture closure (following Meindl and Lovejoy, 1985) was the primary method used, with confirmatory evidence of age seriation obtained from the presence of endocranial impressions from arachnoid granulations (also known as

Pacchionian granulations). Individuals were then categorised as Young Adult (18 to 25 years), Middle Adult (25 to 40 years), or Older Adult (>40 years) in order to facilitate further analysis.

Sex was determined on the basis of secondary sex characteristics as exhibited in the morphology of the skull (Buikstra and Ubelaker, 1994) and the size of teeth, and the individuals were then characterised as Female or Male. The particulars of age and sex for each pathological condition vary according to the number of teeth and jaws that can be observed for analysis, and therefore are presented in tabular detail only for the conditions reported. Overall, however, the sample consisted of 72 males, 45 females and six individuals of indeterminate sex. Middle aged adults comprised over half of the sample when it was partitioned by age; this is probably due to a combination of the difficulty in obtaining precise age estimates for adult skeletons and the typical life expectancy in antiquity, and is not thought to cause problems for this analysis.

All identifiable teeth and fragments of the maxillae and mandibles were examined, inventoried, and their pathological lesions scored according to accepted disciplinary standards (Hillson, 1996). Teeth that were too fragmentary to identify were not included in the study, and those that were unobservable for particular features were omitted from certain analyses. Inter- and intra-observer error in the assessment of pathological lesions was assessed through re-scoring a portion of the sample and was found to be negligible.

A total of 1077 teeth were observable for pathological lesions (27% of the sample size that would be expected if all 123 individuals had possessed a full set of teeth). Occlusal macrowear was scored for all teeth following the illustration and accompanying descriptors developed by Smith (1984). Smith's eight stages of wear were then collapsed into four stages for further analysis: none (Stage 1), slight (Stages 2-4), moderate (Stages 5-6), and severe (Stages 7-8). This categorisation served to increase the subsample sizes for comparison, but, more importantly, took into account those cases where the observer felt that the wear was difficult to assign to only one stage and noted during recording that the wear was, for example, '5 to 6'.

A total of 1259 alveoli were observable for antemortem and postmortem loss (32% of the expected sample size). Antemortem tooth loss was recorded in cases of missing teeth when the alveolus exhibited a reactive process of bone resorption and/or deposition (i.e., healing); in cases of missing teeth where the alveolus was unremodelled the tooth was assumed to have been lost postmortem.

In addition, macroscopically observable lesions on the glenoid fossa (also known as the mandibular fossa) and the articular eminence (tubercle) of the temporal bone, and on the mandibular condyles, were scored according to the descriptors developed for changes to the form and surface of the condyle (Wedel, Carlsson and Sagne, 1978; Richards and Brown, 1981). Form change ranges from no observable change through slight and marked remodelling (including lipping at the anterior margin of the joint and clear presence of cortical cysts) to deforming change; surface change ranges from no change through uneven and irregular surface (the latter accompanied by perforation of the compact bone), to destruction of the compact bone that is greater than 3 mm^2 in size. Radiographs were not employed in this study. Not all of the individuals presented complete temporomandibular joints (see table 1) but either mandibular condyles or glenoid fossae (and sometimes both) were observable in 116 individuals out of the total sample of 123 individuals.

The associations of AMTL and occlusal wear with disorders of the TMJ were examined statistically but required some modifications to categories of tooth loss and wear data: these factors are normally scored per tooth, whereas evidence of TMJ alteration is normally scored by individual. Thus, the average tooth loss score and the average occlusal macrowear score were calculated per individual, and the student's t-test was used to compare dental scores between individuals affected and unaffected by TMJ alterations.

Intra- and intercemetery analyses of cranial and dental non-metric traits (Johnson and Lovell, 1994; Prowse and Lovell, 1996) have shown that the individuals buried in Cemeteries B and G are not epigenetically distinguishable; therefore the remains of these two cemeteries have been grouped for further analysis. This made it possible to use Fisher's Exact Test, which is applicable to 2 x 2 contingency tables, to compare very small subsamples (where at least one expected value fell below 5). Other statistical tests that were used to identify intra- and inter-cemetery patterns included the chi-squared test and the chi-squared test with Yate's correction for continuity when comparing small subsamples, the latter used when at least one expected value fell below 10.

Results and discussion

Occlusal macrowear

In all three cemeteries, the majority of teeth exhibited slight occlusal wear, with molars displaying the most cases of severe wear (table 2). Indeed, none of the observable incisors and canines exhibited occlusal wear scored as severe. Statistically insignificant differences in the severity of wear were seen when the age categories were compared, although the teeth of younger adults showed markedly less severe wear, and the teeth of older adults in the non-elite cemetery sample more commonly displayed severe wear. The teeth of males tended to have more moderate and severe degrees of occlusal wear than did those of females.

Antemortem tooth loss

Table 3 presents the tooth count and individual count results for AMTL at Naqada. Overall, males and females were found to have comparable rates of AMTL (35% and 33% respectively), and more than 80% of older adults had lost at least one tooth antemortem. Individuals in Cemetery T suffered less AMTL but the difference is not statistically significant; this result may be an artifact of small and unequal samples.

The ultimate cause of AMTL may be difficult to identify: severe tooth wear may lead to dislocation of a tooth and may cause a tooth's anchor in the alveolus to be tenuous. In addition, carious invasion of the pulp chamber of a tooth may lead eventually to granulomata, abscesses, and cysts that lead to loss of alveolar bone in the periapical region. Several individuals in the Naqada sample exhibit wear so severe that the tooth roots functioned in occlusion, but the wear occurred slowly because secondary dentin filled in the pulp chamber as the wear progressed, preventing carious bacteria from entering the tooth and leading to periapical abscess with subsequent tooth loss.

	Cemeteries B & G				Cemetery T			
	None	Slight	Moderate	Severe	None	Slight	Moderate	Severe
Tooth Class								
Incisor	0	0	0	0	0	0	0	0
Canine	0	0	0	0	0	2	0	0
Premolar	0	12	7	5	0	10	5	2
Molar	0	37	26	10	0	51	19	11
Total	0	50	34	16	0	63	24	13
Sex								
Male	0	23	24	10	0	29	19	13
Female	0	23	10	3	0	30	5	0
Total	0	46	34	13	0	59	24	13
Age								
Older Adult	0	12	17	15	0	22	15	6
Middle Adult	0	18	17	3	0	28	14	9
Young Adult	0	13	4	0	0	7	0	0
Total	0	43	38	18	0	56	29	15

Table 2: Patterns of occlusal macrowear at Naqada (tooth count). Percentages have been rounded to the nearest full number. Some totals ≠ 100 due to rounding.

	Cemeteries B & G			Cemetery T			Total		
	n	N	%	n	N	%	n	N	%
Teeth	159	1070	15	21	189	11	180	1259	14
Individuals	38	104	37	5	25	20	43	129	33

Table 3: Tooth count and individual count patterns of antemortem tooth loss at Naqada, by cemetery. n = number of affected teeth/individuals; N = number of observable teeth/individuals; % = n/N x 100.

Lesions of the Temporomandibular Joint

The age, sex, and cemetery distribution of individuals affected by lesions of the temporomandibular joint is presented in table 4. The overall prevalence of individuals affected by lesions of the TMJ is 31%, which is slightly smaller than frequencies reported for other ancient Egyptian samples (Leek, 1972) and Mesolithic Nubia (Greene, 1972). In all but one case the porosity, osteophyte development, and changes in surface morphology found on the mandibular condyles, glenoid fossa, and eminence are consistent with a diagnosis of osteoarthritis (OA). Figures 1, 2, and 3 show examples of the TMJ alterations that have been classified as representative of OA.

A noninflammatory disorder, OA is characterised by 1) destruction of the disc that cushions the articulation of the mandibular condyle with the glenoid fossa of the temporal bone, 2) resorption and proliferation of bone on the mandibular condyle and the glenoid fossa, and 3) lesions of the convex articular eminence that lies anterior to the glenoid fossa. The precise cause of TMJ lesions is difficult to ascertain, in large part because the anatomy of the TMJ is very complex, a combination hinge and sliding joint that is further complicated by the presence of a fibrocartilaginous disc rather than a hyaline cartilage disc, and by the relationship between the joint and the dentition. Despite these unique characteristics, the TMJ shares features with other synovial joints, such as a fibrous joint capsule, connective ligaments, and lubrication by synovial fluid.

	Cemeteries B & G			Cemetery T			Total		
	n	N	%	n	N	%	n	N	%
Age & Sex									
Female									
Young adult	0	7	0	0	1	0	0	8	0
Middle adult	2	8	25	3	7	43	5	15	33
Older adult	5	14	36	0	4	0	5	18	2
Male									
Young adult	2	7	29	0	0	0	2	7	29
Middle adult	8	25	32	2	6	33	10	31	32
Older adult	10	26	39	4	11	36	14	37	38
Total	27	87	31	9	29	31	36	117	31

Table 4: The age, sex, and cemetery distributions of TMJ lesions at Naqada (individual count). n = number of affected individuals; N = total number of individuals; % = n/N x 100. Percentages have been rounded to the nearest full number.

The frequency of TMJ lesions increases with age, for both sexes and both cemetery groups, but very small subsamples prevented a statistical assessment of the effects of age on the prevalence of TMJ lesions. Although advanced age is not a primary cause of TMJ lesions, it is considered a predisposing factor because of the cumulative effects of loading of the joint and the tendency of the disc to become irreparably fatigued and weakened. The relationship between the joint and the dentition is altered by malocclusion, tooth wear, and AMTL. Loss of vertical height occurs with aging as the occlusal surfaces of teeth become worn, and this loss of height can be a significant factor in altered biomechanics at the TMJ. Vertical height often is maintained, however, by overeruption of the teeth involved, which is made possible by the deposition of cementum at the root apices. (This phenomenon can be seen in skeletal remains as the cemento-enamel junction becomes more distant from the alveolar margin, but it must be differentiated from the resorption of the alveolar margin that occurs due to periodontal disease). The early antemortem loss of a tooth, with the continued eruption of the opposing tooth, or the dislocation of adjacent teeth, also can affect the biomechanics of chewing. Sheridan and co-workers (1991) found a highly significant relationship between long-term tooth loss and TMJ disorders in Medieval Nubians. In response to such changing biomechanical stresses, the TMJ remodels as it attempts to maintain the form and function of the joint.

Regrettably, the incomplete nature of most of the dentitions from Naqada made a detailed assessment of the relationship between malocclusion and/or occlusal wear and TMJ lesions difficult. Indeed, an examination of the individuals in which TMJ lesions were exhibited shows that 18 individuals with OA lesions in one or both glenoid fossae did not have an associated mandible. In general, AMTL and severe occlusal wear were not associated with TMJ lesions, but, this intrapopulational pattern may be deceiving and a more focused assessment of individual cases may prove to be informative in this regard. Perhaps illuminating is the fact that three of the four individuals with what was classified as severe OA lesions at the glenoid fossae had occlusal wear of the first molars that was scored as stage 7 or 8 (i.e., severe). Although Sheridan and co-workers (1991) did not find a correspondence between occlusal wear and TMJ lesions among Medieval Nubians, Hodges (1991) observed a clear association between these variables in a series of ancient British skeletal remains.

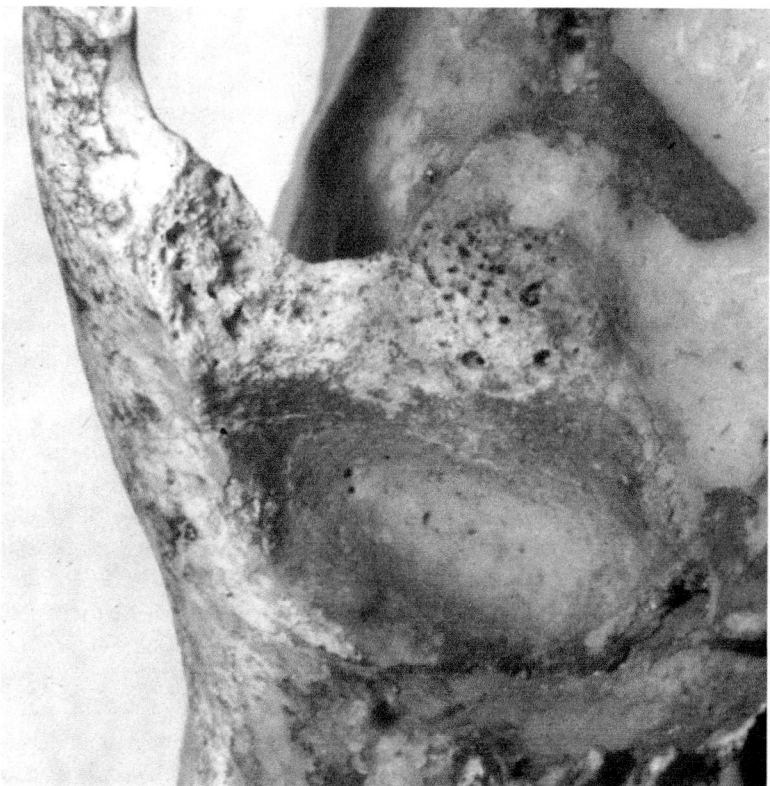

Figure 1: Resorption on the right articular eminence
(basilar view of the cranium), classified as a 'slight' degree of osteoarthritis.

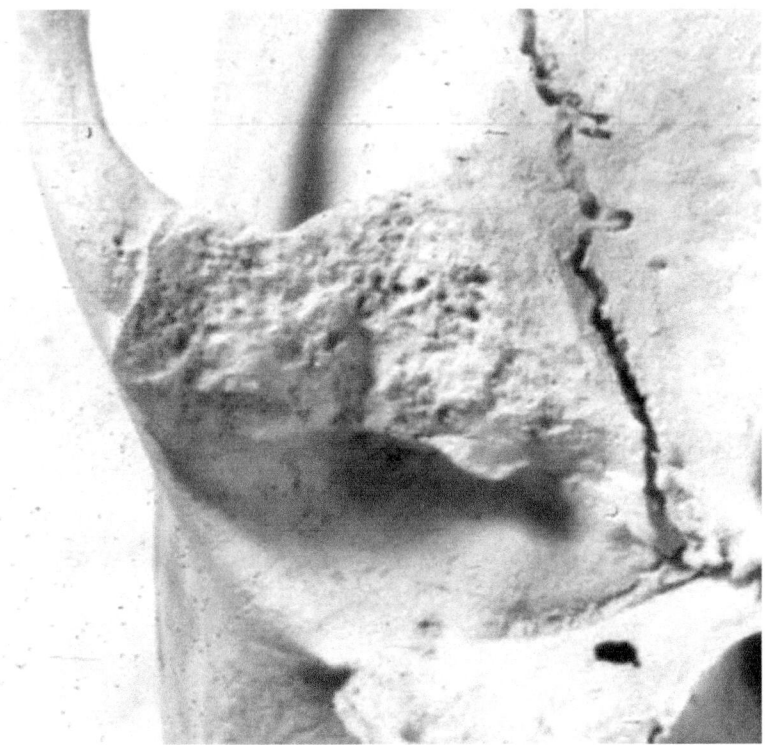

Figure 2: Extensive resorption and marginal lipping on the right articular eminence, classified as 'moderate' osteoarthritis.

Males and females in the young adult category could not be compared because of missing values, but when the sexes in other age classes are compared there is no statistically significant difference in the frequency of TMJ lesions when the cemetery is controlled.

As with the results obtained for an analysis of sex differences in the frequency of TMJ lesions, there is no statistically significant difference between the frequency of TMJ lesions among the individuals buried in the elite versus non-elite cemeteries. When severity of TMJ arthritis is considered, however, three of four individuals with severe lesions came from Cemetery B/G and only one was from Cemetery T, a pattern that cannot be explained by age differences. While it might be supposed that differences would exist because of differences in diet, i.e., with non-elites consuming a diet that included coarser foods, any dietary difference was not translated into observable differences in the overall pattern of osteoarthritis at the TMJ. There are (at least) three explanations for this: 1) social stratification, and hence dietary difference, was not as pronounced during the ancient Egyptian Predynastic period as it was during later periods of pharaonic rule; 2) the individuals buried at Naqada were exposed to other causes of TMJ arthritis, such as parafunctional activities like bruxism, regardless of their status; and 3) the causes of TMJ lesions may have been different for elite and non-elite individuals, but the end result was the same.

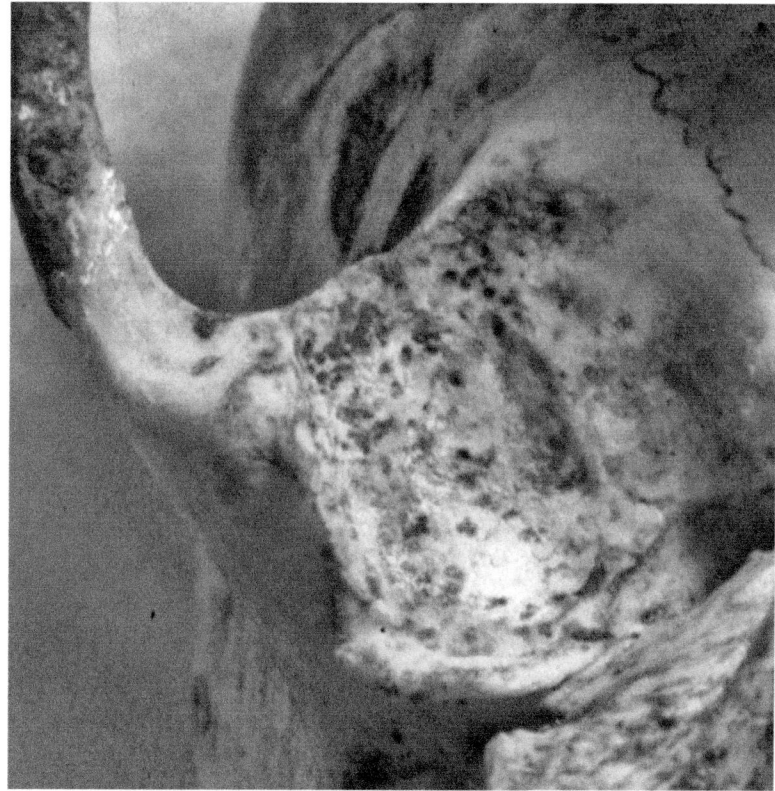

Figure 3: Pronounced remodelling of the right glenoid fossa with the formation of grooves and eburnation, classified as 'severe' osteoarthritis.

One unusual display of TMJ lesions was found in an individual from Cemetery T. Regrettably, the associated mandible was not available for examination, but both glenoid fossae exhibit what appear to be incipient ankylosis resulting from ossifications within the joint that were broken repeatedly during movement of the mandibular condyles (figure 4). A number of potential causes of these lesions were considered; it is widely reported in the clinical literature that TMJ ankylosis is most commonly a unilateral occurrence and usually is due to trauma to the mandible, infection (from the middle ear and mastoid), or a congenital condition such as bifid mandibular condyle. Without a mandible to examine, the identification of a definitive diagnosis is beyond the scope of this paper, but an inflammatory form of arthritis or a pathological condition affecting the fibrocartilaginous disc must be considered in the differential diagnosis.

Conclusions

No statistically significant differences in the prevalence of TMJ arthritis were detected when males and females were compared. Nor were differences detected between the elite and non-elite cemetery samples. The frequency of TMJ lesions increases with age, for both sexes and both cemetery groups, but this association was not statistically significant.

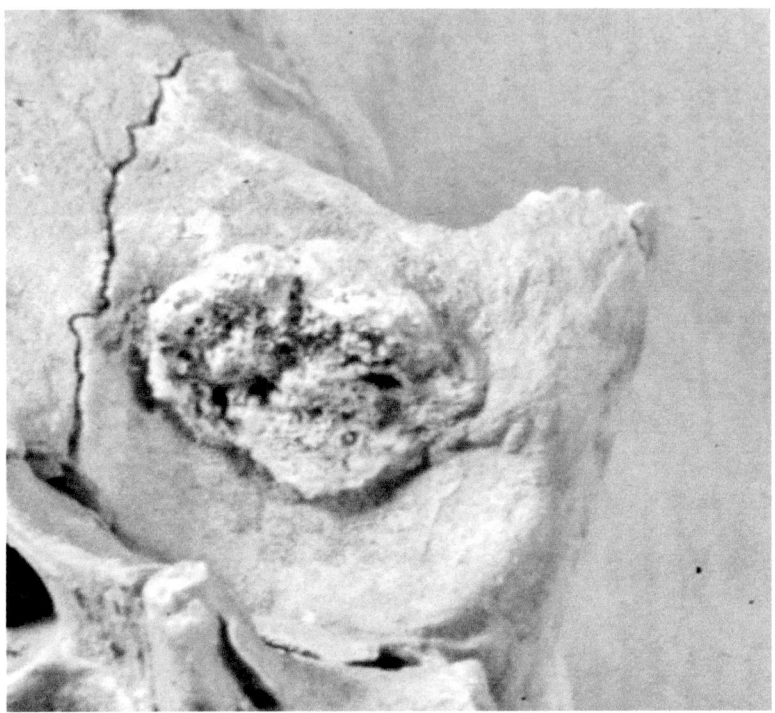

Figure 4: Proliferation of bone within the left glenoid fossa.

Regrettably, the incomplete nature of most of the dentitions from Naqada made a detailed assessment of the relationship between malocclusion and/or tooth wear and TMJ lesions difficult. In general, however, AMTL and occlusal wear were not associated with TMJ lesions: comparisons of average AMTL frequencies and of average occlusal wear scores between individuals with and without TMJ arthritis revealed no statistically significant differences. Since occlusal wear is severe in many individuals, however, it seems likely that a reduction in vertical occlusal height affected TMJ function, and a more focused assessment of individual cases may prove to be informative in this regard.

Acknowledgements

I thank the Social Sciences and Humanities Research Council of Canada and the University of Alberta for financial support of this research; and, for their assistance, Marnie Bartell (University of Alberta); and Robert Foley, Corinne Duhig, and Maggie Bellatti (University of Cambridge).

References

Bartell, M., 1994. *Palaeopathology of Cranial Remains from Predynastic Naqada, Egypt.* MA Thesis, Edmonton, Alberta: University of Alberta.

Brothwell, D.R., 1963. The macroscopic dental pathology of some earlier human populations. In D. R. Brothwell (ed.), *Dental Anthropology*. London: Pergamon Press. pp.272-287.

Buikstra, J.E. and Ubelaker, D.H. eds., 1994. *Standards for Data Collection from Human Skeletal Remains*. Research Series No. 44, Fayetteville, Arkansas: Arkansas Archaeological Survey.

Fawcett, C.D. and Lee, A.C., 1902. A second study of the variation and correlation of the human skull, with special reference to the Naqada crania. *Biometrika*, 1, pp.408-467.

Forshaw, R., 2009. Dental health and disease in ancient Egypt. *British Dental Journal*, 206, pp.421-424.

Greene, D.L., 1972. Dental anthropology of early Egypt and Nubia. *Journal of Human Evolution*, 1, pp.315-324.

Greene, T.R., 2006. *Diet and Dental Health in Predynastic Egypt: A Comparison of Hierakonpolis and Naqada*. PhD thesis, Fairbanks, Alaska: University of Alaska Fairbanks.

Hillson, S., 1996. *Dental Anthropology*. Cambridge: Cambridge University Press.

Hodges, D.C., 1991. Temporomandibular joint osteoarthritis in a British skeletal population. *American Journal of Physical Anthropology*, 85, pp.367-377.

Johnson, A.L. and Lovell, N.C., 1994. Biological differentiation at Predynastic Naqada, Egypt: An analysis of dental morphological traits. *American Journal of Physical Anthropology*, 93, pp.427-433.

Keita, S.O.Y., 1990. Studies of ancient crania from northern Africa. *American Journal of Physical Anthropology*, 83, pp.35-48.

Keita, S.O.Y. and Boyce, A.J., 2001. Diachronic patterns of dental hypoplasias and vault porosities during the Predynastic in the Naqada region, Upper Egypt. *American Journal of Human Biology*, 13, pp.733-743.

Leek, F.F., 1972. Bite, attrition and associated oral conditions as seen in ancient Egyptian skulls. *Journal of Human Evolution*, 1, pp.289-295.

Meindl, R.S. and Lovejoy, C.O., 1985. Ectocranial suture closure: a revised method for the determination of skeletal age at death based on the lateral-anterior sutures. *American Journal of Physical Anthropology*, 68, pp.57-66.

Miller, J., 2008. *An Appraisal of the Skulls and Dentition of Ancient Egyptians, Highlighting the Pathology and Speculating on the Influence of Diet and Environment*. BAR International Series 1794, Oxford: Archaeopress.

Petrie, W.M.F. and Quibell, J.E., 1896. *Naqada and Ballas 1895*. London: Bernard Quaritch.

Prowse, T.L. and Lovell, N.C., 1996. Concordance of cranial and dental non-metric traits and evidence for endogamy in ancient Egypt. *American Journal of Physical Anthropology*, 101, pp.237-246.

Richards, L.C. and Brown, T., 1981. Dental attrition and degenerative arthritis of the temporomandibular joint. *Journal of Oral Rehabilitation*, 8, pp.293-307.

Sheridan, S.G., Mittler, D.M., van Gerven, D.P. and Covert, H.H., 1991. Biomechanical association of dental and temporomandibular pathology in a medieval Nubian population. *American Journal of Physical Anthropology*, 85, pp.201-205.

Smith, B.H., 1984. Patterns of molar wear in hunter-gatherers and agriculturalists. *American Journal of Physical Anthropology*, 63, pp.39-56.

Warren, E., 1897. An investigation on the variability of the human skeleton: with special reference to the Naqada race. *Philosophical Transactions of the Royal Society B*, 189, pp.135-227.

Wedel, A., Carlsson, G.E. and Sagne, S., 1978. Temporomandibular joint morphology in medieval skull material. *Swedish Dental Journal*, 2, pp.177-187.

How to make a mummy:
A late hieratic guide from Abusir

Jiří Janák and Renata Landgráfová

Czech Institute of Egyptology, Charles University in Prague, Czech Republic

Abstract

In the course of the exploration of the shaft tomb of Menekhibnekau at Abusir, one of the largest preserved embalmer's deposits was discovered in 2003. The deposit contained, besides a few other things, over 300 large amphorae. Twenty of the large amphorae and 30 of the smaller jars bore short inscriptions in late hieratic or demotic script, one imported Phoenician amphora bears several different texts in Phoenician and Aramaic. The late hieratic texts identify a number of embalming materials and in some cases connect these with a day-number. As these go beyond 'day 60', they cannot be days of the month, but represent, given the context, the days of the mummification process. The embalmer's deposit of Menekhibnekau thus contains a unique 'cookbook' for making mummies.

Introduction

Ancient Egyptian mummies have fascinated the Western world since the 16th century (Raaven and Taconis, 2005, p.19), but our knowledge of the mummification process is still imperfect. It relies on a number of sources of various types and reliability. The most famous of these is Herodotus' account of the Egyptian mummification process in Book II of his Histories, which mentions that the most elaborate mummification process took about 70 days (Histories II.86-88). Contracts, bills, lists etc. from mummification workshops contain information about the substances that were needed for the mummification process to be completed (e.g. Ikram and Dodson, 1998, p.105). Two Egyptian texts, the Rituel d'Embaumement (Sauneron, 1952) and the Embalming Ritual of the Apis (Vos, 1993) give a number of interesting details on the mummification process, but rather than representing embalmer's handbooks, they concentrate on the religious rituals connected with mummification and on the actions that accompany the mummification ritual in the divine sphere. Recently, the most useful information has come from experiments, both on human (Brier and Wade, 1999) and animal bodies (Ikram, 2005, pp.18-43), as well as the research projects of The University of Manchester (e.g. David, 2008) and other academic and research institutions. In many details, however, we are still left in the dark. Hopefully, the current volume with its numerous contributions will help shed some more light on the mummification process.

In the course of the exploration of the shaft tomb of Menekhibnekau (a high-ranking official who lived at the turn of the 26th and 27th Dynasties) at Abusir, one of the largest preserved embalmer's deposits was discovered in 2003 (Bareš, Smoláriková and Strouhal, 2005; Bareš, Janák, Landgráfová and Smoláriková, 2010). The deposit (marked as Shaft

Vessel	Day	Text	Translation
XVII		n3 ʿrf.w p3 tms	bags, red linen
CIV		n3 ʿrf.w	bags
XLVI	36	n3 ʿrf.w	bags
CLXIII		n3 ʿrf.w	bags
XCI		[p3] tms	red linen
LXXXV		p3 tms	red linen
LIII	24	p3 tms hnʿ n3 ʿrf.w	red linen and bags
CCXXV	32	p3 tms hnʿ n3 ʿrf.w	red linen and bags
LX	24	p3 tms hnʿ n3 ʿrf.w	red linen and bags
LXII		[p3] tms hnʿ [ʿr]f.w	red linen and [bag]s
XXII		p3 tms n n3 ms.w ḥr.w	red linen of the Children of Horus
XCIII		p3 tms n n3 ms.w ḥr.w	red linen of the Children of Horus
XVIII		p3 tms n n3 ms.w ḥr.w	red linen of the Children of Horus
LXXXVIII		p3 [ḥ]smn n n3 ms.w ḥr.w	natron of the Children of Horus
CCXIX		p3 ḥs[mn ...]	nat[ron ...]
CCLXIV		[p]3 ḥsmn	natron
CCLXXXII		pḥr.t ḥnk.t mḥ-1	beer potion, the first
XCIII drop	52	gs.w	balm
XCII drop b		dj.t r jwf [...]	bandages [...]
CLXXIX drop	36	dj(.t) r jwf db3	bandages, db3 cloth
LXVIII drop b	28	dj(.t) r jwf db3 mn[ḥ.t]	bandages, db3- and mnḥ.t cloth
CLI cooking	40	dj(.t) r jwf db3 mnḥ.t	bandages, db3- and mnḥ.t cloth
CXXVI drop	44	dj(t) r j[wf db3] mnḥ.t	banda[ges, db3-] and mnḥ.t cloth
CXXXV drop a	45	dj(.t) r jwf db3 mnḥ.t	bandages, db3- and mnḥ.t cloth
LXXXVII drop	52	dj(.t) [r]jwf db3 m[nḥ.t]	bandages, db3- and m[nḥ.t] cloth
CLXIX drop	32	dj(.t) r jwf=f sndm stj=f	to be placed on his body to sweeten its smell
XXV drop		ʿntjw sfj psj	myrrh and cooked resin
XCII drop a		sfj	resin
LXVI cooking		sfj	resin
CCCII drop		sft	resin
CCCXX cooking b		sft ʿntj[w ...]	resin, myrr[h ...]
XVIII drop	63	sfj ʿntjw	resin, myrrh
CVIII drop		sfr ʿntjw w3d	resin, fresh myrrh
LXVIII drop a		sft ʿntjw w3d	resin, fresh myrrh
CLXIII drop		sft ʿntjw w3d	resin, fresh myrrh
LXXXV drop	60	sfj ʿntjw w3d	resin, fresh myrrh
XXX beaker		ḥsmn ʿntjw	natron, myrrh
CCCXX cooking		p3 ʿntjw n n3 ḥnk.t	myrrh of (or for) beer
XXX drop	60	snfr w3d n[d]m	green eye paint, ointment

Table 1: Summary of late hieratic inscriptions from the embalmer's cache of Menekhibnekau. The large amphorae are indicated by grey background.

S1) has an inverted E-shaped ground plan and contained, besides a few other things (cf. Smoláriková, 2011), over 300 large amphorae. These contained mainly sand and straw, although some of these large amphorae contained one or more smaller vessels – drop-shaped jars, beakers and cooking-pots – some broken, some still intact. Twenty of the large amphorae and 30 of the smaller jars bore short inscriptions in late hieratic or demotic script, one imported Phoenician amphora bears several different texts in Phoenician and Aramaic (for this text see Dušek and Mynářová, 2011). It is, however, the late hieratic corpus of inscriptions that concerns us here, as it is directly connected with the mummification process. The texts identify a number of embalming materials, and in some cases, connect these with a day-number. As these go beyond 'day 60', they cannot be days of the month, but represent, given the context, the days of the mummification process. The embalmer's deposit of Menekhibnekau thus contains a unique 'cookbook' for making mummies.

Inscriptions from the embalmer´s deposit

The exploration of the embalmer's deposit, including the careful recording of each individual vessel's exact location and contents, lasted several years, and only in the course of the spring excavation season of 2010 were the last of the vessels stored within the embalmer's cache extracted, examined and documented. The analysis of inscriptions proceeded in parallel to the archaeological work, with each new inscription being carefully analysed in the context of the ones already known. The inscriptions appear in two formats. The first type only identifies one or more materials used in the mummification process (it can be presumed that these materials were brought to the mummification workshop in these vessels). In the second inscription type, the materials are either preceded or followed by an identification of the day of the mummification process on which they were to be used, in the form *hrw mh*-NUMBER. As the texts are short, repetitive and similar, the best way to present them here is in tabular form (see table 1).

Table 1 summarises all of the late hieratic inscriptions from the embalmer's cache of Menekhibnekau. The Roman numerals without qualification refer to the large amphorae, those with qualification ('cooking' for cooking pot, 'drop' for drop-shaped jar and 'beaker' for beaker) refer to the smaller vessels found within them. Further qualification by a letter means that a single amphora contained more than one smaller vessel. When the inscription includes a day-date, this is noted in the second column. Before assessing the importance of these texts for our knowledge of the mummification process, some of the less straightforward transcriptions and translations need to be treated in more detail.

Red linen and bags

For the inscriptions mentioning 'red linen' and 'bags', often together, that of Amphora LIII can be taken as a representative example:

	hrw mh-24	24th day
	p3 tms hnʿ	red linen and
	n3 ʿrf.w	bags

Of the two words that identify the contents of the vessels, *n3 ꜥrf.w* is the easier one to interpret. Wörterbuch (Erman and Grapow, 1971, I, p.210) translates ꜥrf with 'Beutel, Säckchen', i.e. 'bag', and the context of mummification makes one immediately think of the bags with natron that were used to fill the cavities of the body in order to help its desiccation. For the use of these bags, cf. Ikram, 2005, pp.29-43; Brier and Wade, 1997; 1999. For examples found in an embalmers' cache, see Winlock, 1941, pl. III. The second word, *p3 ṯms*, is more difficult. The Wörterbuch (V, 369-370) gives two root meanings: ⟨hieroglyphs⟩, ⟨hieroglyphs⟩ or ⟨hieroglyphs⟩ 'red', 'red colour' and ⟨hieroglyphs⟩ 'evil'. The determinatives used for *p3 ṯms* on the amphorae differ greatly, however, given that most of the identifiable determinatives are either ⟨hieroglyph⟩ or ⟨hieroglyph⟩ or a combination of both, the word will most likely derive from the core meaning 'red'. We had originally thought of 'red pigment,' which appears in an embalmer's receipt from the Ptolemaic period (Ikram and Dodson, 1998, p.105), but neither the large quantities nor the timing (days 24 and 32 of the mummification process) are easily explicable for pigment. A possible explanation is suggested by the text on amphorae LIII and LXII, where *p3 ṯms* has the determinative ⟨hieroglyph⟩, which indicates that the word refers to a type of cloth. This is compatible with the quantities and earlier stage of the mummification process, and we have therefore adopted the translation 'red linen' for *p3 ṯms*. There are attestations of the use of red linen – referred to with different words – in the course of the embalming ritual of the Apis bull (e.g. Quack, 1995, p.127).

If such a reading is proven to be correct, it could shed a new light on the main stage of the embalming process. The above-shown close link between the *ṯms* cloth and the bags with natron is particularly noteworthy, since it shows that both materials were used on the same day of mummification, or even simultaneously. Studies on the embalming process, as well as experiments, have shown that such bags or packets filled with natron should be placed into the thoracic and abdominal cavities (Brier and Wade, 1997, p.94) and then (probably more than once) replaced by new packages with fresh, clean natron (Ikram, 2005, pp.16-43). This would suggest that the *ṯms* cloth must have been used in a similar and contemporaneous embalming procedure as the natron bags.

Moreover, 'a linen with small balls wrapped in it' was found in Weni's burial in Abydos, and this piece of cloth was interpreted as 'a linen that had been used to dry out the body' (Lacovara and Richards, 2011; also personal communication). A different interpretation is, however, possible. Judging from the above-mentioned information about the context in which the *ṯms* cloth is attested, and in the light of problems faced during recent experiments with the embalming process (Ikram, 2005, pp.29-43), it seems more probable that this cloth was placed over the corpse before loose natron was applied over it. Such an interpretation explains why this cloth was mentioned several times together with the natron bags, as it would simplify the embalming process, especially the cleaning of the body, and to prevent the loose natron from sticking to the skin or accidentally falling into body cavities (for these difficulties see Ikram, 2005, pp.31-43).

Amphora XCIII, with its fullest spelling, can be used as representative of the second type of inscription, mentioning 'red linen of the Children of Horus':

p3 ṯms n n3 ms.w ḥr.w
red linen of the Children of Horus

The translation, 'red linen of the Children of Horus,' can be derived from all three attestations; the reading *ms.w* becomes clear from the full spelling of amphora XCIII, reproduced here in hieroglyphs. Unfortunately, no day-date is clearly associated with any of the *ms.w ḥr.w* texts. It is, however, reasonably certain that they refer to the treatment of the organs to be placed in the canopic jars, guarded by the four sons of Horus (cf. Sauneron, 1952, 4 [pBoulaq III, II, 16–17]; on the Children of Horus in the context of mummification cf. also Smith and Dawson, 1924 pp.63-4.)

With regard to the above presented interpretation of the *ṯms* cloth ('red linen'), it is important that this type of cloth is mentioned again in connection to the preservation of the body (in this case, of the internal organs). Again, we suggest that the red linen in question was placed over the organs ready to be embalmed, or these organs might have been actually wrapped in the cloth, before applying loose natron over them. Such use of the cloth probably had similar effects on the embalmed internal organs as have already been suggested for the mummification of the body, mainly the reduction of natron's destructive effect that must have been essential for the preservation of the internal organs. Natron used to desiccate the internal organs was stored in amphora LXXXVIII, labelled 'natron of the Children of Horus'.

Beer and balm

Amphora CCLXXXII bears a unique text in the corpus, mentioning a 'beer potion' accompanied by the ordinal numeral 'the first':

pḫr.t ḥnk.t mḥ-1
beer potion, the first

Most likely, the 'beer potion' or 'beer remedy' was a liquid that was used to wash out the cavities of the mummy-in-making (Cf. Goyon, 1972, p.32 for a wine mixture used for this purpose, and Erman and Grapow, 1971, I, 549, 12 and Quack, 1999, 28-30 for *pḫr.t* used in the mummification process). Here it is identified as 'the first'. This may mean that there were more such potions to be used to cleanse the body from within, and this particular one was the first of these to be employed, or, perhaps more likely, this is to be read as 'the first (day)'. In this case, the reference would be to the primary cleansing of the body from within immediately after its arrival at the mummification workshop and the removal of its internal organs, which had to be done quickly to prevent deterioration processes from starting in the mummy-to-be.

With the beer potion, we leave the large amphorae and move onto the smaller vessels. First, there is the simple 'balm' of drop-shaped jar XCIII, identified as belonging to the 52nd day of the mummification process:

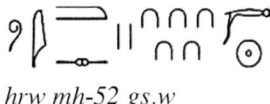

hrw mḥ-52 gs.w
52nd day, balm

The *gs*-balm appears several times in the embalming ritual (Sauneron, 1952), cf. pBoulaq III, 2,5 and 2,17 (*gs ms.w ḥr.w*, 'balm of the Children of Horus', in the latter case). The fact that balm is connected with the 52nd day shows in itself that the body could not have spent 70 days in natron, but that a significant part of the 70-day period was taken up by the actual embalming and wrapping of the body.

Bandages

Wrapping is what the following eight texts indicate, as they mention two different types of cloth. Interestingly, all but one (which is, however, incomplete and may also have contained a day-date) include indications of the day on which they were to be used. The text of drop-shaped jar CXXXV can serve as an example of these texts:

hrw mḥ-45 dj(.t) r jwf ḏbȝ mnḫ.t
45th day, bandages, *ḏbȝ*- and *mnḫ.t* cloth

The reading of the first element, *dj.t r jwf*, 'bandages' (lit. 'to be placed on the flesh') is established by the text of drop-shaped jar XCIIb, where the infinitive *dj.t* is written out in full, and by the different text of drop-shaped jar CLXIX, where an unmentioned substance is 'to be placed to his flesh to sweeten its smell' (and connected with the 32nd day). In the other cases, the cloths are 'to be placed on the flesh' – i.e. to wrap the mummy. The whole phrase could hypothetically mean either 'bandages and *ḏbȝ*- and *mnḫ.t* cloth' or 'bandages of *ḏbȝ*- and *mnḫ.t* cloth'. Given the size of the vessels, the latter is far more likely (we are leaving the translation ambiguous, in accordance with the Egyptian original). The days of the mummification process connected with the bandages are 28, 40, 44, 45 and 52. It is also worth mentioning that *mnḫ.t*-cloth is the cloth that is most frequently used in the Rituel d'Embaumement to wrap the individual parts of the mummy, cf. pBoulaq III, 2,19 and 2,23 (*mnḫ.t šps.t* used to cover the mummy) and pBoulaq III, 4,9-4,10 (wrapping the head of the mummy with *mnḫ.t*-cloths of various deities).

The early day 28 – as opposed to days 40 and later – might be explained either by using the cloths (otherwise used as bandages) for cleaning the body during the desiccation process at the times when wet natron was exchanged for fresh, or they could be bandages used for wrapping the internal organs that had been dried separately and probably took less time to desiccate (we owe the latter idea to Ryan Metcalfe).

Resin, myrrh and eye paint

The next group of inscriptions is concerned with resin and myrrh, and the two days connected with them are day 60 and 63. We can use the text of drop-shaped jar LXXXV as representative for these inscriptions:

sfj ʿntjw w3ḏ   resin, fresh myrrh
hrw mḥ-60   60th day

The first of the substances, resin, appears most frequently as *sfj*, but also as *sfr* and even *sft*; it is, however, in all these cases a single substance (for the possible spellings of resin, cf. Koura, 1999, pp.177-180). Myrrh, *ʿntjw*, is spelt consistently, and in some cases qualified by the adjective *w3ḏ*, 'green, fresh'. Interestingly, resin is in one case, that of the drop-shaped jar XXV, qualified by the verb *psj* (written with the sign Q7), 'to cook', i.e. cooked resin as opposed to fresh myrrh. There is a Demotic parallel to this text in P. Brooklyn 35.1462: *sfj jw=f ps* (Vittmann, 2006).

The next text appears only on cooking pot CCCXX and is somewhat unclear:

p3 ʿntjw n n3 ḥnḳ.t
myrrh of (or for) beer (plural)

The expression 'myrrh of the beer (plural)' may refer to myrrh used in or in connection with the beer potion whilst washing the body cavities. There is also a reference to myrrh used together with beer (or even within the brewing process) in a cultic text from Ptolemaic Period: 'Take for yourself the wonderful beer, which the noble one has brewed with her hands, with the beautiful plant from Geb and myrrh from Nepy.' (Edfou I, 2, 367: 9–11).

Finally, the inscription of drop-shaped jar XXX connects eye paint and ointment with the 60th day of the mummification process:

snfr w3ḏ n[ḏ]m   green eye paint, ointment
hrw mḥ-60   60th day

Conclusions

Now that we have analysed the individual texts and identified the substances mentioned, we can concentrate on the day-dated inscriptions, summarised below in table 2. This should reveal the approximate sequence of the use of the individual substances within the mummification process.

Vessel	Day	Text	Translation
LIII	24	p3 tms hnꜥ n3 ꜥrf.w	red linen and bags
LX	24	p3 tms hnꜥ n3 ꜥrf.w	red linen and bags
LXVIII drop b	28	dj(.t) r jwf db3 mn[h.t]	bandages, db3- and mnh.t cloth
CCXXV	32	p3 tms hnꜥ n3 ꜥrf.w	red linen and bags
CLXIX drop	32	dj(.t) r jwf=f sndm stj=f	to be placed on his body to make his smell pleasant
XLVI	36	n3 ꜥrf.w	bags
CLXXIX drop	36	dj(.t) r jwf db3	bandages, db3 cloth
CLI cooking	40	dj(.t) r jwf db3 mnh.t	bandages, db3- and mnh.t cloth
CXXVI drop	44	dj(t) r j[wf db3] mnh.t	banda[ges, db3-] and mnh.t cloth
CXXXV drop a	45	dj(.t) r jwf db3 mnh.t	bandages, db3- and mnh.t cloth
LXXXVII drop	52	dj(.t) [r]jwf db3 m[nh.t]	bandages, db3- and m[nh.t] cloth
XCIII drop	52	gs.w	balm
LXXXV drop	60	sfj ꜥntjw w3d	resin, fresh myrrh
XXX drop	60	snfr w3d n[d]m	green eye paint, ointment
XVIII drop	63	sfj ꜥntjw	resin, myrrh

Table 2: Substances sorted by days of the mummification process.

The table shows quite interesting results. Red linen and bags (of natron) appear in use on day 24 and 36 of the process of mummification – we may presume these were the times when the natron filling of the mummy was exchanged. By day 32, the mummy had probably exuded a rather bad smell, so that it became necessary to use the contents of drop-shaped jar CLXIX, described as 'to be placed on his flesh to make its smell pleasant'. On day 36, bandages and cloths start being employed in larger amounts; this continues until day 52 of the mummification process. Finally, between days 52 and 63, balms, resins, myrrh, ointment and green eye paint are all being employed. Is this picture, approximate as it must be given that we only have a partial record, consistent with what other sources tell us about the mummification process in the Late Period?

It is interesting to compare our results with textual evidence from the Late Period and Graeco-Roman papyri and stelae collected by Maria Cannata during her recent research on the topic of the burial in Late Period and Graeco-Roman Egypt (Cannata, 2007; also personal communication). The results of this comparison are given below in table 3.

The entries marked in bold in table 3 are those where the evidence of both sources matches. Given the numerous overlaps, we can probably safely presume that the changes in the mummification process between the Saite-Persian period and the Graeco-Roman period were not very significant. Unfortunately, M. Cannata's results are of no help in searching for a solution for our 'beer potion' problem, as it is unlikely that the embalmers would postpone the evisceration of the body to day 16, which is where wine comes in her list. Perhaps, as the substance called *sfj* comes consistently at an early stage in the Graeco-Roman textual sources, we should interpret it as '(fragrant) oil', in which case the combination of oil and cloths could be used for the same purpose as our beer mixture, the washing of the body. Somewhere in the early days of the mummification process, the body was immersed in natron for the first time, and the natron would be exchanged, probably twice, on day 24 and 32 of the mummification process. On day 36 (Menekhibnekau; the Graeco-Roman sources give day 35), wrapping began, using vast numbers of cloths. The final touches on the mummy began with applying balm on day 52 (both source types agree here) and resins and myrrh on days 60 and 63, with the burial ensuing on day 70 (according to the Graeco-Roman sources, any time between day 64 and 70 is possible for Menekhibnekau).

day	M. Cannata	Menekhibnekau S1
1	*sjf*-resin, *ḥbs*-cloth, *mnḫ*-cloth	beer potion (???)
4	cloths, *ks*	
16	*jp*-wine, *sjf*-resin, *sbn*-bandages, cloth	
24		red linen and bags (with natron)
28	*sjf*-resin, **cloth**	*db3*- and *mnḥ.t* **cloth**
32		red linen and bags (with natron); (something) to be placed on the mummy to make it smell pleasant
35	**cloths; wrapping begins?**	
36		*db3*- and *mnḥ.t* **cloth**
40		*db3*- and *mnḥ.t* **cloth**
44		*db3*- and *mnḥ.t* **cloth**
45		*db3*- and *mnḥ.t* **cloth**
52	cooking of *mtḥ*-**unguents**, **cloths**	*gs.w*-**balm**; *db3*- and *mnḥ.t* **cloth**
60		resin, fresh myrrh; green eye paint, ointment
63		resin, myrrh
70	*ḥb-ks*	

Table 3: Comparison of the evidence of texts from Shaft S1 of the tomb of Menekhibnekau and Graeco-Roman demotic papyri and stelae analysed by M. Cannata.

Thanks to the combined evidence of the Late and Graeco-Roman textual sources and the inscriptions on the vessels in the embalmers' cache of the tomb of Menemkhibnekau, we are now able to assign more precise timings to the individual processes that went on in a mummification workshop of the Late Period onwards. Hopefully, this will shed a new or even crucial light into the mysteries of the process of embalming.

References

Bareš, L., Smoláriková, K. and Strouhal, E., 2005. The Saite-Persian Cemetery at Abusir in 2003. *Zeitschrift für Ägyptische Sprache und Altertumskunde* 132, pp.95-106.

Bareš, L., Janák, J., Landgráfová, R. and Smoláriková, K., 2010. The Shaft Tomb of Menekhibnekau at Abusir – Season of 2008. *Zeitschrift für Ägyptische Sprache und Altertumskunde* 137 (2), pp.91-97.

Brier, B. and Wade, R.S., 1997. The Use of Natron in Human Mummification: A Modern Experiment. *Zeitschrift für Ägyptische Sprache und Altertumskunde* 124, pp.89–100.

Brier, B. and Wade, R.S., 1999. Surgical procedures during ancient Egyptian mummification. *Zeitschrift für Ägyptische Sprache und Altertumskunde* 126, pp.89 - 97.

Cannata, M., 2007. Of Bodies and Soles. The Meaning of the Root qs in the Ptolemaic Period. In: M. Cannata, ed. *Current Research in Egyptology 2006. Proceedings of the Seventh Annual Symposium.* Oxford: Oxbow, pp.21–42.

David, R., 2008. *Egyptian Mummies and modern science.* Cambridge: Cambridge University Press.

Dušek, J. and Mynářová, J., 2011. Phoenician and Aramaic inscriptions on a Phoenician storage jar (Excav. no. 826/S/10): A preliminary report. In: L. Bareš and K. Smoláriková, eds. *The Shaft Tomb of Menekhibnekau.* Volume I. Prague: Czech Institute of Egyptology, Faculty of Arts, Charles University in Prague. pp.179-181.

Erman, A. and Grapow, H., 1971. *Wörterbuch der Ägyptischen Sprache.* Berlin: Akademie Verlag.

Goyon, J.C., 1972. *Rituels funéraires de l'ancien Égypte.* Paris: Cerf.

Herodotus, unknown date, *The Histories.* Translated by J.M. Marincola and A. De Selincourt., 1996. London: Penguin Classics.

Ikram, S., 2005. *Divine Creatures. Animal Mummies in Ancient Egypt.* Cairo and New York: The American University in Cairo Press.

Ikram, S., 2010. Mummification. In: J. Dieleman and W. Wendrich, eds. *UCLA Encyclopedia of Egyptology,* Los Angeles. [online] Available at: <http://escholarship.org/uc/item/0gn7x3ff> [Accessed 2013]

Ikram, S. and Dodson, A. 1998. *The Mummy in Ancient Egypt. Equipping the Dead for Eternity.* London: The American University in Cairo Press.

Janák, J. and Landgráfová, R. 2011. Texts from the embalmers' deposit. In: L. Bareš and K. Smoláriková, eds. *The Shaft Tomb of Menekhibnekau. Volume I: Archaeology.* Prague: Czech Institute of Egyptology, Faculty of Arts, Charles University in Prague. pp.164–178.

Koura, B., 1999. *Die '7-Heiligen Öle' und andere Öl- und Fettnamen* [Aegyptiaca Monasteriensia 2], Aachaen: Shaker Verlag.

Lacovara, J. and Richards, P., 2011. *iAbydos* (web-site and blog). [online] Available at: <http://iabydos.wordpress.com/> [Accessed 2013].

Quack, J.F., 1995. Zwei Handbücher der Mummifizierung im Balsamierungsritual des Apisstieres, *Enchoria,* 22, pp.123-129.

Quack, J.F., 1999. Balsamierung und Totengericht im Papyrus Insinger. *Enchoria,* 25, pp.27-38.

Raaven, M.R. and Taconis, W.K., 2005. *Egyptian Mummies. Radiological Atlas of the Collections in the national Museum of Antiquities in Leiden.* Turnhout: Brepols.

Sauneron, S., 1952. *Rituel de l'embaumement: Pap. Boulaq III, Pap. Louvre 5.158.* Cairo: Imprimerie Nationale.

Smith, G.E. and Dawson W.R. 1924. *Egyptian Mummies.* London: George Allen and Unwin.

Smoláriková, K., 2011. The embalmers' deposit. In: L. Bareš and K. Smoláriková, eds. *The Shaft Tomb of Menekhibnekau. Volume I: Archaeology.* Prague: Czech Institute of Egyptology, Faculty of Arts, Charles University in Prague. pp.81–163.

Vittmann, G., 2006. P. Brooklyn 35.1462. *Enchoria,* 30, pp.155–160.
Vos, R.L., 1993. *The Apis Embalming Ritual P. Vindob. 3873* [OLA 50]. Louvain: Peeters Press.
Winlock, H.E., 1941. *Material Used at the Embalming of King Tutankhamun.* New York: Metropolitan Museum of Art.

Studying mummies: Giving life to a dry subject

Michael R. Zimmerman

Villanova University, Philadelphia, USA

Abstract

Paleopathology, the study of disease in ancient remains, adds the crucial dimension of time to improve our understanding of the evolution of diseases and their role in human biological and social history. While information on ancient disease patterns can be obtained from historical records and such diverse media as paintings, pottery effigies, figurines and religious statuary, the examination of skeletal material and mummies yields the most reliable palaeopathological information.

How do we study mummies?

Egyptian mummies can be examined by standard postmortem studies and hold a great potential for palaeopathological examination. The preservation of tissues postmortem is based on inactivation of destructive enzymes in invading bacteria and fungi and in the tissues themselves. In the modern laboratory this process of fixation is accomplished by immersion of the tissue in a liquid chemical fixative, usually some variant of formaldehyde. Rapid fixation results in good preservation. In contrast, mummification is often erratic and anything but rapid, classical sources describing a 70 day period of mummification in ancient Egypt. The inevitable result is a degree of tissue destruction that may obscure pathological change or even underlying tissue structure.

While the loss of the features of colour and consistency, so useful in examining fresh tissue, limits the interpretation of gross pathologic change, the tissues can be rehydrated and sections prepared for microscopic examination. Rehydration of desiccated tissue is based on the use of a solution developed in early 20th century by Sir Marc Armand Ruffer, the father of modern palaeopathology (Ruffer, 1921).

Diagnoses of many conditions can be made with a considerable degree of confidence and accuracy. The examination of mummified bodies requires the collaborative efforts of specialists from many fields. Radiographic techniques tell much about bodies before they are removed from their coffins or have been unwrapped (Harris and Weeks, 1973; Harris and Wente, 1980). Computed tomography (CT) scanning (Rühli et al., 2002; O'Brien et al., 2009)) and magnetic resonance imaging (Rühli et al., 2007) allow for more thorough examination of mummies either before or instead of traditional autopsy procedures and can be followed by endoscopically guided biopsies of sites of interest, thereby avoiding the often somewhat destructive nature of a full autopsy examination. Light and electron microscopy can demonstrate remarkable preservation of normal and diseased structures.

Ruffer's eponymous solution of water, alcohol and sodium carbonate is effective in rehydrating mummified tissue, although poorly preserved tissue tends simply to dissolve. Rehydrated specimens are sectioned and stained according to standard histologic procedures. A variety of special stains can be used to demonstrate specific features of the tissues. In general, the connective tissues and any foreign elements, such as pigments, bacteria, or parasites are best preserved.

As in all other branches of scientific and medical investigation, the study of mummies will be facilitated by new technology. The past century has seen the progression of radiological study from relatively basic X-rays to sophisticated CT analysis, resulting in non-destructive examinations, including CT guided endoscopic biopsies, yielding a high level of diagnoses.

Enhancements in nuclear magnetic resonance technology have allowed the examination of mummies without the need for rehydration. Gas chromatography mass spectrometry has been used in the study of ancient Egyptian embalming materials. Future advances in palaeoserology may allow the detection of antibodies to pathogenic microorganisms, while improvements in DNA detection are already expanding our knowledge of relationships and the history and evolution of disease. The analysis of mummified material for ancient DNA (aDNA) has received much recent interest (Paabo et al. 1988; Paabo, 1985; Cano, et al., 2000; Haak et al., 2002) and a major meeting was held for 2013 (Royal Society, London, 2013). Although there are technical problems related to the survival of aDNA and contamination by modern DNA, applications can include the determination of genetic relationships, such as the remarkable recent clarification of the relationships among the members of King Tutankhamun's family (Hawass et al., 2010). There are technical problems related to the survival of aDNA and contamination by modern DNA. I was involved in a DNA study of evidence related to the assassination of President Kennedy, but contamination resulted in an unsuccessful effort (Zimmerman, 2007).

What mummies do we study?

For most people, mummies mean those from Egypt, studies of which have been undertaken since the 19th century. However, mummies are found in many areas of the world. The oldest human mummies, of the Chinchorro culture from what is now northern Chile and southern Peru, date to 5000 to 3000 BCE, thousands of years before Egyptian mummies (Arriaza, 1995). Frozen bodies, buried by accident or rarely deliberately, are found in the Arctic (Beattie, 1987; Hart Hansen and Gullov, 1989; Notman et al., 1986; Paddock et al., 1970; Smith and Zimmerman, 1975; Zimmerman, 1985; Zimmerman and Aufderheide, 1984; Zimmerman and Smith, 1975; Zimmerman and Tedford, 1976; Zimmerman et al., 1981; Zimmerman et al., 1971) and of course there is the famous 5300 year old Iceman, found in the Alps between Austria and Italy 20 years ago (Spindler, 1994; Gostner and Vigl, 2002; Cano et al., 2000).

What have we learned from mummies?

The examination of mummies has two goals: fitting the diagnosis of diseases in individual mummies into a picture of the health status of a given ancient population; and providing information on the evolution of diseases. Diseases fitting into a number of broad general categories diagnosed in mummies have approached both of these goals, although many diagnoses offered in these studies of mummies have been presumptive, a necessary limitation considering the poor preservation often encountered (Zimmerman, 2001; 2004; 2012).

A number of congenital skeletal deformities and traumatic injuries have been seen in mummies. Clubfoot has recently been diagnosed in Amenhotep III and Tutankhamun, which would have resulted in difficulty in walking, which accounts for numerous depictions of Tutankhamun being seated in activities that normally require an upright posture, as well as the presence of numerous walking sticks, several showing wear, in Tutankhamun's tomb (Hawass et al., 2010). A mummy from the Dakhleh Oasis suffered from a club foot and his liver showed the scarring of cirrhosis, perhaps due to self-medication by an excess of wine during his life (Zimmerman and Aufderheide, 2010).

A rare condition, alkaptonuric arthritis, was diagnosed radiologically in several mummies in the 1960s (Wells and Maxwell, 1962). This hereditary disorder results in a distinctive calcification of the intervertebral cartilages. It has become apparent that the increased radiodensity of these cartilages seen in many Egyptian mummies is pseudopathology, an artefact of mummification rather than a real pathological condition. Fractures have been noted in Egyptian mummies, some of which appear to have been 'embalmer's fractures', caused by fitting a mummy into an undersized coffin.

One of the most notable traumatic injuries was diagnosed in the Iceman by CT scanning. An arrowhead was identified in the left shoulder region, which resulted in a fatal haemorrhage (Gostner and Egarter, 2002).

There have been many reports of infectious and inflammatory processes. Infectious disease is categorised by the etiological agents. Viruses are below the range of light microscopy and have not been seen in the few electron microscopic studies performed on mummies. X-ray examination of the Pharaoh Siptah showed an overall shortening of the entire right leg and atrophy of the soft tissues, indicating the presence of a neuromuscular disease in childhood, diagnosed by some as characteristic of poliomyelitis (Harris and Weeks, 1973) although by others as cerebral palsy (Ikram and Dodson, 1998). A smallpox-like eruption has been noted on the mummy of Ramses V (Ruffer, 1921).

One of the most common bacterial infections and a major cause of death in the pre-antibiotic era is pneumonia, which has been diagnosed in Egyptian mummies. Tuberculosis has been well documented in ancient Egypt (Morse et al., 1964) and pre-Columbian South America (Allison et al., 1973). The mummy of an Egyptian child showed spinal curvature due to tuberculosis with death due to pulmonary haemorrhage (Zimmerman, 1979). As the pathogenesis of tuberculosis in a young child is infection from a much older adult, this finding confirms the archaeological evidence of extended multigenerational households in ancient Egypt, a pattern that persists today. Of interest is the absence of evidence of tuberculosis in predynastic Nubian skeletons and mummies, suggesting the dynastic period for the onset of human tuberculosis in the Nile Valley, but recent molecular evidence indicates a much older date for the evolution of human tuberculosis (Stone et al., 2009).

For protozoan parasites our evidence is mostly indirect. Ruffer detected enlarged spleens in two Egyptian mummies and diagnosed malaria. DNA studies have determined that Tutankhamun suffered from falciparum malaria, the most serious form of the disease (Hawass et al., 2010). Parasitic worms and their ova remain well preserved for millennia, and the characteristic ova of *Ascaris lumbricoides*, *Schistosoma hematobium*, and *Taenia solium* have been reported in Egyptian mummies (Reyman et al., 1977; Ruffer, 1910; Lambert-Zazulak, 2003; Contis and David, 1996; Reyman et al., 1977; Jonckheere, 1944). The Dakhleh Oasis is far from the Nile but a mummy showed the characteristic ova of schistosomiasis - probable evidence of a trade route via oases in the western desert (Zimmerman and Aufderheide, 2010). Cysticercosis, a complication of pork tapeworm infestation, has been seen in an Egyptian mummy (Reyman et al., 1977).

The skin is often well preserved and careful examination can be rewarding. A rare skin condition, subcorneal pustular dermatosis, was reported in a 3200 year old Egyptian mummy (Zimmerman and Clark, 1976). This condition was not described in modern patients until 1956 by two British physicians, Doctors Sneddon and Wilkinson. One of the lessons of palaeopathology is that diseases may exist long before their modern clinical diagnosis.

Dental and middle ear disease have also long been part of the human condition. Periodontal disease and caries have been noted in pharaohs and fellahs. These conditions can lead to infection of the middle ear and mastoid sinuses, and perforated eardrums have been seen in an Egyptian mummy (Benitez, 1988).

The degenerative process most commonly seen in mummies is osteoarthritis, often seen in Egyptian mummies, where its presence in the hot dry climate of Egypt and Nubia belies the folk attribution of the disease to damp climates. X-rays of Ramses II have revealed severe osteoarthritis in his hips (Harris and Wente, 1980).

A more life-threatening disorder, atherosclerosis, has been very well documented by historic evidence and the finding of atherosclerosis in many Egyptian mummies, including the Pharaoh Merneptah. A more recent CT study identified the disorder in 9 of 22 mummies in the Cairo Museum (Allam et al., 2009). This much higher incidence than had previously been reported raises the question of the cause of this disease. Ancient Egyptians did not smoke cigarettes, eat much meat or sugar, or deal with the environmental pollution or stresses of the 21st century. One theory of the cause of this disorder is that it is actually an infectious disease, caused by as yet unidentified bacteria, analogous to the recent discovery of the infectious cause of stomach ulcers. Perhaps those bacteria have been waging this 'arms race' for thousands of years.

Diagnosis of the various disturbances of circulation, such as myocardial infarction ('heart attack') resulting from atherosclerosis, has not been made in any mummy. This is surprising in view of the historical and anatomical evidence of atherosclerosis noted above, but appears to be the result of a problem in preservation. Necrotic heart muscle is autolysed *in situ*, and as such would not be distinguishable from post-mortem autolysis (Zimmerman, 1978; 1993).

Another common degenerative process is the accumulation of foreign material, particularly in the lungs. The combination of carbon and silica particles (anthracosilicosis) has been seen in the lungs of almost all Egyptian mummies. The findings of anthracosis are attributed to life-long exposure to open fires, for heating and cooking, while the silicosis is probably due to inhalation during the sandstorms common to Egypt. Several mummies have shown damage to the lungs as a result of these exposures.

There have only been a few tissue diagnoses of tumours in mummified remains. Benign tumours in Egyptian mummies include a few skin tumours (Horne, 1986; Zimmerman, 1981; Sandison, 1967; Allison and Gerszten, 1982; Fulcheri, 1987), a sacral nerve sheath tumour (Strouhal et al., 2003), and uterine leiomyomas (Strouhal, 1976; Kramar et al., 1983). The diagnosis of malignant tumours in Egyptian mummies is rare. At this point the literature contains only three reports of microscopically confirmed cancer in Egyptian mummies. Cancers of the rectum and of the urinary bladder were found in ca. 200 C.E. mummies from the Dakhleh Oasis, Egypt (Zimmerman and Aufderheide, 2010) and a rare tumour of skeletal muscle (rhabdomyosarcoma) was reported in the mummy of a Peruvian child (Gerszten and Allison, 1991).

It has been suggested that the short life span of individuals in antiquity precluded the development of cancer. Although this statistical construct is true, individuals in ancient Egypt did live long enough to develop such diseases as atherosclerosis and osteoporosis. It must also be remembered that in modern populations, bone tumours primarily affect the young. Another explanation for the lack of tumours in ancient remains is that tumours might not be well preserved but my experimental study of mummification indicates that the features of malignant cells are favourable to preservation by mummification. In an ancient society lacking surgical intervention, evidence of cancer should remain in all cases. The virtual absence of malignancies must be interpreted as indicating their rarity in antiquity and indicate that carcinogenic factors are limited to societies affected by modern industrialization and tobacco usage (David and Zimmerman, 2010).

What will we learn from mummies?

As in all other branches of scientific and medical investigation, the study of mummies will be facilitated by the development of new technology. The past century has seen the progression of radiological study from relatively basic X-rays to sophisticated CT analyses, resulting in non-destructive examinations, including CT guided endoscopic biopsies, yielding a high level of diagnoses.

Future advances in palaeoserology may allow the detection of antibodies to pathogenic microorganisms, while improvements in DNA detection are already expanding our knowledge of the history and evolution of diseases. Enhancements in nuclear magnetic resonance technology have allowed the examination of mummies without the need for rehydration. Gas chromatography mass spectrometry has been used in the study of ancient Egyptian embalming materials. Such advances can also facilitate a look back at studies in the past. The University of Manchester's KNH Centre for Biomedical Egyptology and The Natural History Museum, London, have begun a re-examination of the more than 7000 specimens found during the first Archaeological Survey of Nubia.

In a 21st century computerised world where privacy has been literally 'blown to bits' (Abelson et al., 2008), caution has been raised regarding issues of privacy for ancient historical figures. Should they have the same rights as deceased modern individuals? Should rules be developed for mummy studies, changing our approach to threatening diseases such as cancer, as well as our understanding of the past? Palaeopathological study of mummies expands our knowledge of the life stories and fate of ancient individuals, their relationship to others, ancient migrations, the evolution of disease, and the role of ancient disease in human evolution and social history, as well as applications to modern medicine and implications for health in the modern world.

References

Abelson, H., Ledeen, K. and Lewis, H., 2008. *Blown to Bits: Your Life, Liberty and Happiness after the Digital Explosion.* Boston: Pearson Education.

Allam, A.H., Thompson, R.C., Wann, L.S., Miyamoto, M.I., and Thomas, G.S., 2009. Computed tomographic assessment of atherosclerosis in ancient Egyptian mummies. *Journal of the American Medical Association*, 302, pp.2091-2094

Allison, M.J, and Gerszten, E., 1982. *Paleopathology in South American mummies.* Richmond, VA: Virginia Commonwealth University.

Allison, M.J., and Gerszten, E., 1983. Case no. 13: Hydronephrosis of the kidney, trauma, homicide, decapitation, abdominal laceration. *Paleopathology Club Newsletter*, 16, p.1.

Allison, M.J., Mendoza, D. and Pezzia, A., 1973. Documentation of a case of tuberculosis in pre-Columbian America. *American Review of Respiratory Disease*, 107, pp.985-991.

Arriaza, B., 1995. *Beyond Death: The Chinchorro Mummies of Ancient Chile.* Washington, Smithsonian Institution Press.

Aufderheide, A.C., 2003. *The Scientific Study of Mummies.* Cambridge: Cambridge University Press.

Beattie, O. and Geiger, J., 1987. *Frozen in Time: Unlocking the secrets of the Franklin expedition.* Saskatoon: Western Producer Prairie Books.

Benitez, J.T., 1988. Otopathology of Egyptian mummy PUM II: final report. *Journal of Laryngology and Otology*, 102, pp.485-490.

Cano, R.J., Tieffenbrunner, F., Ubaldi, M., Del Cueto, C., Luciani, S., Cox, T., Orkand, P., Künzel, K.H. and Rollo, F., 2000. Sequence analysis of bacterial DNA in the colon and stomach of the Tyrolean Iceman. *American Journal of Physical Anthropology*, 112, pp.297-309.

Contis, G. and David, A.R. 1996. The epidemiology of bilharzia in Ancient Egypt: 5000 Years of Schistosomiasis. *Parasitology Today*, 12 (7), pp.253-255.

David, A.R. and Zimmerman, M.R., 2010. Cancer: A new disease, an old disease, or something in between? *Nature Reviews Cancer*, 10, pp.728-733.

Fulcheri, E., 1987. Case no. 27: Verruca vulgaris. *Paleopathology Club Newsletter*, 31, p.2.

Gerzsten, E. and Allison, M., 1991. Human soft tissues in paleopathology. In Ortner, D. C. and Aufderheide, A. C., eds. *Human Paleopathology: Current Syntheses and Future Options*. Washington DC: Smithsonian Press. pp.257-260.

Gostner, P. and Egarter, V.E., 2002. Report of radiological-forensic findings on the Iceman. *Journal of Archaeological Science*, 29, pp.323–326.

Haak, W., Gruber, P., Ruhli, F.J., Boni, T., Ulrich-Bochsler, S., Frauendorf, E., Burger, J. and Alt, K.W., 2002. Molecular evidence of HLA-B27 in a historical case of ankylosing spondylitis. *Arthritis and Rheumatology,* 52 (10), pp.3318-3319.

Harris, J.E. and Weeks, K., 1973. *X-Raying the Pharaohs*. New York: Scribner's.

Harris, J.E., and Wente, E.F., 1980. *An X-ray atlas of the Royal mummies*. Chicago: Chicago University Press.

Hart, G.D. ed, 1983. *Disease in Ancient Man: An International Symposium*. Toronto: Clarke, Irwin & Co.

Hart Hansen, J.P. and Gullov, H.C. eds., 1989. *The Mummies from Qilakitsoq - Eskimos in the 15th Century*. Copenhagen: Nyt Nordisk Forlag

Hawass, Z., Gad, Y.Z., Ismail, S., Khairat, R., Fathalla, D., Hasan, N., Ahmed, A., Elleithy, H., Ball, M., Gaballah, F., Wasef, S., Fateen, M., Amer, H., Gostner, P., Selim, A., Zink, A. and Pusch, C., 2010 Ancestry and pathology in King Tutankhamun's family. *Journal of the American Medical Association*, 303, pp.638-647.

Horne, P.D., 1986. Case no. 23: angiokeratoma circumscriptum. *Paleopathology Club Newsletter*, 27, p.1.

Ikram, S. and Dodson, A., 1998. *Mummies in Ancient Egypt*. London: Thames and Hudson.

Jonckheere, F., 1944. *Une Maladie Egyptienne*. Brussels: Fond AegyptRei Elisa.

Kramar, C., Baud, C.A., and Lagier, R., 1983. Presumed calcified leiomyoma: Morphologic and chemical studies of a calcified mass dating from the Neolithic period. *Archives of Pathology and Laboratory Medicine*, 107, pp.91–93.

Lambert-Zazulak, P., Rutherford, P. and David, A.R., 2003. The International Ancient Egyptian Mummy Tissue Bank at the Manchester Museum as a resource for the palaeoepidemiological study of Schistosomiasis. *World Archaeology*, 35 (2), pp.223-240.

Morse, D., Brothwell, D. and Ucko, P.J., 1964. Tuberculosis in ancient Egypt. *American Review of Respiratory Disease*, 90, pp.524-541.

Notman, D.N.H., Tashjian, J., Aufderheide, A.C., Cass, O.W., Shane, O.C., Berquist, T.H., Gray, J.E. and Gedgaudas, E., 1986. Modern imaging and endoscopic biopsy techniques in Egyptian mummies. *American Journal of Roentgenology*, 146, pp.93-96.

O'Brien, J.J., Battista, J.J., Romagnoli, C. and Chhem, R.H., 2009. CT imaging of human mummies: A critical review of the literature (1979–2005). *International Journal of Osteoarchaeology*, 19, pp.90-98.

Pääbo, S., 1985. Molecular cloning of ancient Egyptian mummy DNA. *Nature*, 314, pp.644-645.

Pääbo, S., Gifford, J.A. and Wilson, A.C., 1988. Mitochondrial DNA sequences from a 7,000 year old brain. *Nucleic Acids Research*, 16, pp.9775-9787.

Paddock, F.K., Loomis, C.C. and Perkons, A.K., 1970. An inquest on the death of Charles Francis Hall. *New England Journal of Medicine*, 282, pp.784-786.

Reyman, T.A., Zimmerman, M.R. and Lewin, P.K., 1977. Autopsy of an Egyptian Mummy: Histopathologic Investigation. *Canadian Medical Association Journal*, 117, pp.470-472.

Royal Society, London 2013. *Ancient DNA: the first three decades*. Available online at: <http://royalsociety.org/events/2013/ancient-dna/>.

Ruffer, M.A., 1910. Note on the presence of *Bilharzia haematobia* in Egyptian mummies of the XXth Dynasty (1250-1000 BC). *British Medical Journal*, 2557, p.16.

Ruffer, M.A., 1921. *Studies in the Palaeopathology of Egypt*. R. L. Moodie, ed. Chicago, University of Chicago Press.

Rühli, F.J., Hodler, J. and Böni, T., 2002 CT-guided biopsy: a new diagnostic method for paleopathological research. *American Journal of Physical Anthropology*, 117, pp.272-275.

Rühli, F.J., von Waldburg, H., Nielles-Vallespin, S., Böni, T. and Speier, P., 2007. Clinical magnetic resonance imaging of ancient dry human mummies without rehydration. *Journal of the American Medical Association*, 298, pp.2618-2620.

Sandison, A.T., 1967. Diseases of the skin. In D. Brothwell and A. T. Sandison, eds. *Disease in Antiquity*. Springfield: C. C. Thomas. pp.449–456.

Smith, G.S. and Zimmerman, M.R., 1975. Tattooing found on a 1,600 year old frozen mummified body from St. Lawrence Island, Alaska. *American Antiquities*, 40, pp.434-437.

Spindler K., 1994. *The man in the ice*. London: Weidenfeld and Nicolson.

Strouhal E., 1976. Tumors in the remains of ancient Egyptians. *American Journal of Physical Anthropology*, 45, pp.613–620.

Strouhal, E., Numeakova, A. and Kouba, M., 2003. Paleopathology of Iufaa and other persons found beside his shaft tomb at Abusir (Egypt). *International Journal of Osteoarchaeology*, 13, pp.331–338.

Stone, A.C., Wilbur, A.K., Buikstra, J.E. and Roberts, C.A., 2009. Tuberculosis and leprosy in perspective. *Yearbook of Physical Anthropology*, 52, pp.66-94.

Wells, C. and Maxwell, B.M., 1962. Alkaptonuria in an Egyptian mummy. *British Journal of Radiology*, 35, pp.679-682.

Zimmerman, M.R., 1978. The mummified heart: a problem in medicolegal diagnosis. *Journal of Forensic Science*, 23, pp.750–753.

Zimmerman, M.R., 1979. Pulmonary and Osseous Tuberculosis in an Egyptian Mummy. *Bulletin of the New York Academy of Medicine*, 55, pp.604-608.

Zimmerman, M.R., 1981. A possible histiocytoma in an Egyptian mummy. *Arch Derm*, 117, pp.364-365.

Zimmerman, M.R., 1985. Paleopathology in Alaskan Mummies. *American Scientist*, 73, pp.20-25.

Zimmerman, M.R., 1993. The paleopathology of the cardiovascular system. *Texas Heart Institute Journal*, 20, pp.252-257.

Zimmerman, M.R., 2001. The study of preserved human tissue. In D.R. Brothwell and A.M. Pollard, eds. *Handbook of Archeological Sciences*. Chichester: Wiley. pp.249-257.

Zimmerman, M.R., 2004. Paleopathology and the study of ancient remains. In C.R. Ember and M. Ember, eds. *Encyclopedia of Medical Anthropology: Health and Illness in the World's Cultures*. New York: Kluwer/Plenum. pp.49-58.

Zimmerman, M.R., 2007. *Report to Assassination Records Review Board*. National Archives, Washington, DC, Dec. 14, 1999. M.R. Zimmerman. Excerpted in: Bugliosi, Vincent, Reclaiming History: The Assassination of President John Fitzgerald Kennedy, endnotes. New York: Norton. pp.424-5.

Zimmerman, M.R., 2012. The analysis and interpretation of mummified remains. In A. L. Grauer, ed. *A Companion to Paleopathology*. New York: Wiley-Blackwell. pp.152-169.

Zimmerman, M.R. and Aufderheide, A.C., 1984. The Frozen Family of Utqiagvik: The Autopsy Findings. *Arctic Anthropology*, 21, pp.53-63.

Zimmerman, M.R. and Aufderheide, A.C., 2010. Seven Mummies of the Dakhleh Oasis, Egypt: Seventeen Diagnoses. *Paleopathology Newsletter*. No. 150, pp.16-23.

Zimmerman, M. R. and Clark, W. H, Jr., 1976. A Possible Case of Subcorneal Pustular Dermatosis in an Egyptian Mummy. *Arch Derm*, 112, pp.204-205.

Zimmerman, M.R. and Smith, G.S., 1975. A Probable Case of Accidental Inhumation of 1,600 Years Ago. *Bulletin of the New York Academy of Medicine*, 51, pp.828-837.

Zimmerman, M.R., Trinkaus, E., LeMay, M., Aufderheide, A.C., Reyman, T.A., Marrocco, G.R., Ortel, R.W., Benitez, J.T., Laughlin, W.S., Horne, P.D., Schultes, R.E. and Coughlin, E.A., 1981. The paleopathology of an Aleutian mummy. *Archives of Pathology and Laboratory Medicine*, 105, pp.638-641.

Zimmerman, M.R., Yeatman, G.W., Sprinz, H. and Titterington, W.P., 1971. Examination of an Aleutian Mummy. *Bulletin of the New York Academy of Medicine*, 47, pp.80-103.

Zimmerman, M.R. and Tedford, R.H., 1976. Histologic structures preserved for 21,300 years. *Science*, 194, pp.183-184.

Microstructural analysis of a Predynastic iron meteorite bead

Diane Johnson,[1] Monica M. Grady,[1,2] Tristan Lowe[3] and Joyce Tyldesley[4,5]

[1] Centre for Earth, Planetary, Space and Astronomical Research, The Open University, Milton Keynes, UK.
[2] The Natural History Museum, London, UK.
[3] Henry Moseley X-Ray Imaging Facility, School of Materials, The University of Manchester, Manchester, UK.
[4] Faculty of Life Sciences, The University of Manchester, Manchester, UK.
[5] The Manchester Museum, The University of Manchester, Manchester, UK.

Abstract

The oldest known example of the use of iron in Egypt is in the form of 9 iron beads discovered within tombs 67 and 133 at the Predynastic Gerzeh cemetery (Wainwright, 1912). Subsequent analysis revealed the beads contain significant levels of nickel, leading to identification of the source of this iron as a meteorite (Petrie and Wainwright, 1912). However the origin of nickel rich iron in antiquity has subsequently been the subject of extensive debate. The broader issues regarding iron in ancient Egypt, its appearance, use, perception and text references remain the subjects of debate today.

Our study is an illustration of the non-destructive analysis of an intact artefact, defining a better understanding of its composition, manufacture, and more broadly the presence of iron within ancient Egyptian tombs. Micro x-ray computed tomography (CT), optical imaging, and scanning electron microscopy (SEM) with energy dispersive x-ray spectroscopy (EDS) were applied to reveal its three dimensional structure and chemistry. Small fragments of metallic iron were identified and found to have chemistry and structure consistent with that of a cold worked iron meteorite with 2.4% volume preserved metallic iron. The bead interior was also found to contain a section of woven flax strands used to string the beads.

Introduction

The earliest archaeological evidence for the working of iron in ancient Egypt dates to approximately 600 BCE (Petrie, 1886). The use of metallic iron prior to the Egyptian Iron Age was scarce with the majority, possibly all, of these examples being nickel rich iron exclusively in the form of funerary equipment. A great deal of debate has revolved around the existence and potential use of iron in ancient Egypt, but scientific evidence is scarce. The early examples of iron were originally assumed to be produced from meteorites because of the presence of nickel. Subsequently studies of iron production in the near east have revealed potential man-made origins of nickel rich iron (Piaskowski, 1982; Varoufakis, 1982; Photos, 1989), hence the need to reconsider the origins of this ancient Egyptian iron and associated implications for its interpretation.

Site background

We analysed an iron bead from tomb 67 of the Gerzeh cemetery, circa 3300 BCE. This site excavation was led by archaeologist Gerald Avery Wainwright during 1911-1912 and assisted by colleague Jocelyn Punket Bushe Fox for the British School of Archaeology in Egypt. Gerzeh was an important cemetery because its artefacts were of an unusual type for Lower Egypt; they were subsequently used to define this Predynastic time period as 'Gerzean culture' by Petrie (although this time is now known more commonly as Naqada II) (Petrie, 1939). The excavation of this cemetery recorded 301 tombs in total, 281 being Predynastic burials, the majority of which were undisturbed (Petrie and Wainwright, 1912). Ground conditions did not appear to have supported the preservation of human remains very well, in particular the bones were unstable so no specimens were collected. Although hair was noted on some skulls, preserved organic remains were generally not found at Gerzeh with the occasional exception such as linen. Fortunately this site excavation was well documented; the tomb cards produced are currently preserved in the Petrie Museum of Egyptian Archaeology, University College London.

Tombs 67 and 133 were somewhat unusual in that they were found to contain iron. Tomb 67, one of the earliest tombs on the site, contained 7 iron beads which were strung with beads of other composition including gold, carnelian and agate in two strands, one found across the neck and the other across the waist (the bead analysed by this study was found at the waist). It also contained other unusual objects and materials for this date and site; these included a copper harpoon, a white limestone mace-head and a small ivory vessel in addition to pottery. The positioning of the skull in relation to the body within this tomb was considered by the excavators to be evidence of the deliberate mutilation of the corpse as a grave rite (figure 1) (Wainwright, 1912). Tomb 133 contained many hundreds of beads composed of many different materials, some more common than others but including lapis lazuli, obsidian and iron. This tomb also contained goods from distant locations such as shells from the Red Sea and the Mediterranean, indicative of the lengthy nature of trade routes at this time.

Sample background

The earliest analysis of the iron beads was performed by Prof W. Gowland (Wainwright, 1912). Reporting ferric oxide 78.7% combined with carbonate, 'earthly matter' and water, he concluded that they were originally thin wrought iron plates bent around a rod and underwent rusting in the tomb. Unfortunately no technical details were recorded of the method of analysis, identification of which of the beads was subjected to this analysis or any details of data processing. The two beads found in tomb 133 led Wainwright to the conclusion that both sets of beads most probably had a common origin dating to S.D. 60-63 (Wainwright, 1912). Subsequently one of the Gerzeh beads obtained from Petrie was analysed by Prof C.H. Desch on behalf of the British Association for the Advancement of Science (Desch, 1928). The results appear to have been published as normalised weight percentage values measured from the oxidised bead as 92.5% iron and 7.5% nickel. While Desch gave no description or image of the bead in his report, all the beads appear heavily oxidised in the original site photography (figure 2) (Petrie and Wainwright, 1912).

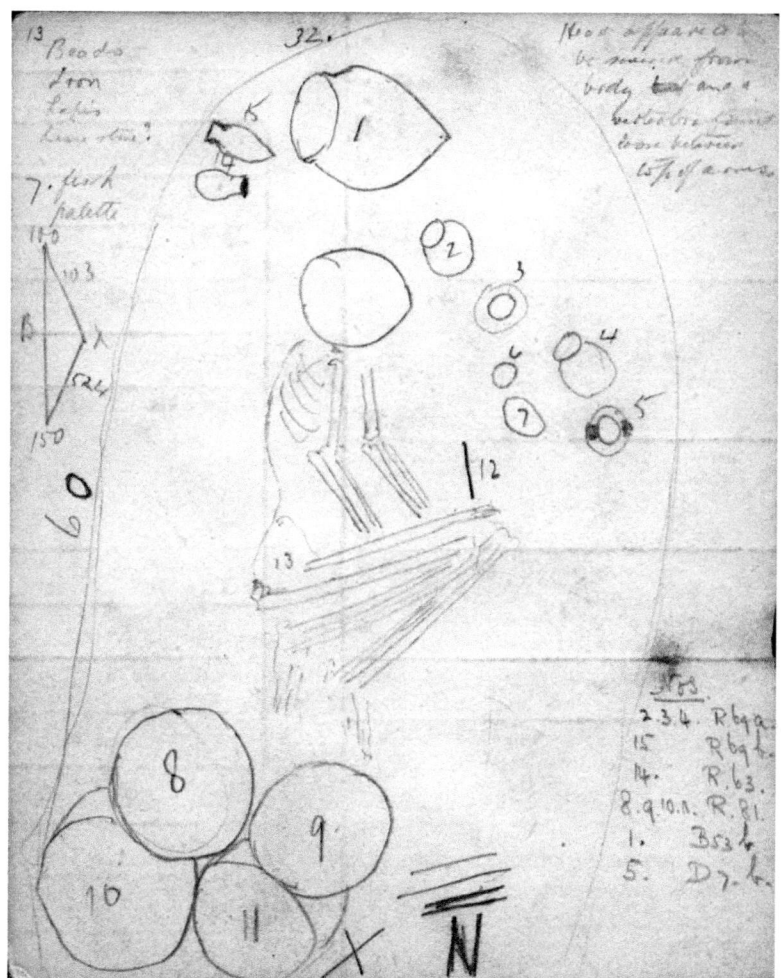

Figure 1: Tomb card of tomb 67 Gerzeh cemetrery
© The Petrie Museum of Egyptian Archaeology, University College London.

Wainwright later published numerous accounts of iron in ancient Egypt. He obviously considered the Gerzeh beads to be celestial in origin citing the analysis result of 7.5 percent nickel as 'proof positive that the iron is meteoric' (Wainwright, 1932). Eventually others proposed new explanations for the origins of nickel rich iron in antiquity within the Near East there is strong evidence to suggest methods including the smelting of nickel rich iron laterites were being practised (Piaskowski, 1982; Varoufakis, 1982; Photos, 1989). However no evidence of this method exists within Egypt or its neighbouring territories at this very early time, but it does prompt a careful reconsideration of all nickel rich iron in antiquity.

The most recent study of a Gerzeh bead employed electron microprobe analysis of surface oxides scraped from the surface of one bead held by the Petrie Museum. This

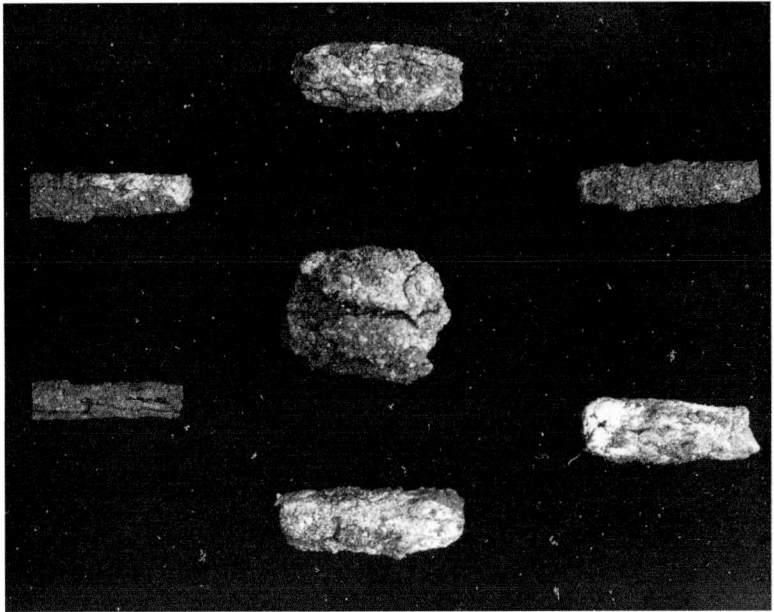

Figure 2: Iron beads found in Gerzeh cemetery tomb 67,
photographed on site the bead analysed in this study is the one displayed lowest on image
(© The Pitt Rivers Museum).

bead had at some point in its past been treated to minimise oxidation of the remaining metal, and the results indicated a maximum of 0.2 wt. % nickel in this surface oxide which was described as the porous hydrated iron oxide, Limonite (El-Gayer, 1995). This study suggested the interpretation of a meteoric origin for the bead iron was weakened by both the low level of nickel recorded in this oxide and the presence of copper up to 0.5 wt. % leading to the suggestion of it being a possible by product of copper smelting. However these low levels of copper could easily be accounted for by contamination, as within tomb 67 a copper harpoon was in relatively close proximity to the beads (Wainwright, 1912) which was not noted by this particular study.

Techniques and methods

In order to determine the true origin of this iron and therefore attempt to understand its related implications, an iron bead from Gerzeh tomb 67 held by The Manchester Museum (accession number 5303) was analysed by scanning electron microscopy (SEM) and micro x-ray computed tomography (CT). Given the nature and rarity of this material, the use of non-destructive methods was important and consideration was given to minimise the risk associated with analysis of this rare material (Richmond, 2005).

SEM as a research tool is frequently applied to gain understanding of surface micro structure and chemistry of materials. Micro x-ray CT is a non-destructive technique used to visualise internal features of an object in 3 dimensions, where x-ray attenuation of the materials within the object define contrast on the resulting recorded image. Egyptology has of course embraced the use of x-ray techniques to study mummified human and animal remains, where frequently CT studies of mummies are so successfully employed that they negate the need for unwrapping these remains. The applications of CT to artefact analysis are less frequent. As no sample preparation is required and the process is non-destructive this can be an ideal technique for research samples requiring high resolution imaging of internal components and structure.

SEM analysis was performed using an FEI Quanta 200 3D SEM with a 0.6nA beam at accelerating voltage of 20kV, with an Oxford instruments 80mm X-Max energy dispersive x-ray detector with Inca software Vs. 4.13. In order to compensate for sample topography geometry hydration and absence of carbon coating data is all expressed as normalised weight percentage. Oxide compositions were recorded over 200µm^2 areas selected to avoid obvious contaminants such as sand grains, point analyses across metal fragmented bands were recorded at 20µm spacing, the data corresponding to metal areas was averaged across four bands. We performed tomography using a Nikon 300kV custom bay recording 2501 projections with an x-ray voltage of 95kV, spot size 3µm producing a model using Avizo®Fire software with a voxel size of 10µm^3. When rendering a 3D model the SEM data was used to identify the compositions of corresponding areas of grey scale contrast of the CT images, semi-quantitatively defining the different components to this complex artefact. Combining these two sets of results allowed a semi-quantitative 3 dimensional model of this bead to be produced.

It was noted upon optical examination that the surface of the bead was highly oxidised and during its oxidation process had undergone alteration by incorporation of grains from the sand with which the tomb had been filled. As can be observed on images recorded on site soon after excavation, and as today, patches of contrast exist caused by areas with an absence of the sand encrusted oxide, probably having separated around the time of excavation. These areas allow direct access to less oxidised areas now exposed as the surface of the bead; these were therefore most suitable for SEM analysis as sites potentially containing preserved metal fragments (figure 3).

Figure 3: The Gerzeh bead analysed in this study shown as it is today, held by The Manchester Museum accession number 5303, scale bar 1cm.

Results

SEM

The outermost surface oxides were consistently found to contain less than 1 wt. % nickel in the form of hydrated iron oxide, Limonite. Figure 4 shows this as clusters of fine blades forming rounded hemispherical structures across the surface especially in protected areas such as the inside edge of the fractures running through the sample. Nickel substitution into this oxide is very minimal, averaging at 0.8 wt. % nickel, other studies have shown minimal nickel substitution into hydroxides such as Lepidocrocite (γ-FeOOH) and Goethite (α-FeOOH) (Golden, Ming and Zolensky, 1995). Areas where the outermost sand encrusted oxides were missing were found to also be highly oxidised but contain preserved metal fragments. The oxides in such areas dominantly exist as Magnetite (Fe_3O_4), having undergone much greater nickel substitution to an average of 4.8 wt. % nickel. Previous studies have found that when iron meteorites weather into oxides, if the main weathering product is magnetite the majority of the nickel from the original metal is retained (Golden, Ming and Zolensky, 1995). Within these areas we find fragments of nickel rich metal, their point analysis averaged nickel content to 27 wt. %. These fragments are elongated and occur in fragmented parallel arrays, this structure and chemistry is as expected for an oxidised, cold worked, octahedral meteorite. At one end of the bead strands of cellular plant fibres with morphology consistent with that of flax were noted (figure 5).

Figure 4: Secondary electron SEM image of hydrated iron oxide growth structures within an open edge of a surface fracture

Figure 5: Secondary electron SEM image of one end of the bead showing exposed fibres of thread originally used to string the beads

CT

The virtual slices produced through both length and width of the bead also show the remains of the thread used to string the beads. The many fractures occurring through the bead are clearly visible, the outer oxide layer with sand grains is identified as are fragmented layers of metals appearing as bright white areas caused by the high x-ray attenuation of metal in contrast to the weathered oxide areas. The CT virtual slice through the width of the bead shows a band of preserved metal with bending points and a feature corresponding to the joining of the two ends of the flat piece of iron which was used to produce this bead. The CT data was used to build a 3D model where the bead components were calculated as 2.4 vol. % metal, 68.6 vol. % nickel rich oxide, 29.0 vol. %, nickel poor oxide (figure 6).

Interpretation

Iron meteorites

Iron meteorites contain very specific chemical and structural signatures that help to define their formation and history within space. There are three fundamental types of meteorites: irons, stones and stony irons. The irons are broadly considered to represent the core of small planets or large asteroids, the stones represent the mantle or crustal materials and the stony irons the core mantle boundary. Within some iron meteorites micro textures are observed. These are 3 dimensional crystallographic growth patterns on a scale large enough to be seen by the naked eye if the surface is etched to reveal the grain boundaries (figure 7). Iron meteorite classification is based upon crystallographic structure and chemistry, they are composed of two nickel rich alloys, one of bulk nickel content approx. 6 wt. % known as the body centred cubic mineral α-Kamacite, the other being the face centred cubic mineral Taenite with 20-65 wt. % nickel (Hutchinson, 2004).

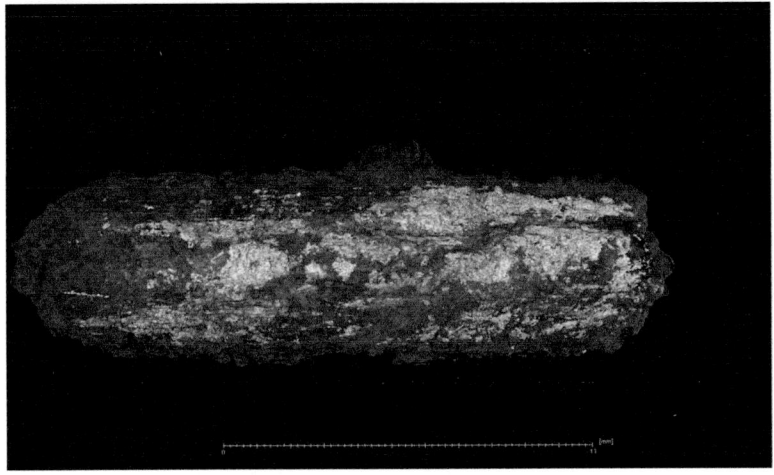

Figure 6: Image of Gerzeh bead CT model oxide layers are represented in semi-transparent form, preserved metal is represented as solid white areas

Commonly meteorites have a bulk metal chemistry content in the range of 5.7 to 16.0 wt. % nickel (Hutchinson, 2004, p.324), on cooling during their formation within a planetary core both alloys grow; initially Taenite structures form with subsequent Kamacite growth parallel to the four octahedral planes inside the γ phase of the Taenite. These are known as octahedral iron meteorites; when cut open, polished and etched these meteorites display parallel linear bands that represent 3 dimensional octahedral crystallographic growth structure, this is known as Widmanstätten pattern (Burke, 1986; Wasson, 1985). The widths of these crystallographic bands are a method of structural classification of these meteorites, a general trend being observed with an increase in nickel content correlating with a decrease in width of kamacite bands (Hutchinson, 2004). The presence of a Widmanstätten pattern within iron-nickel alloys is recognition that the metal is of meteoritic origin.

Experimental iron archaeology

Experimental archaeology has failed to produce layered high-nickel composition forged steel objects, because whenever nickel content exceeds approximately 3 wt. %, the increased brittleness of the metal causes the nickel-rich fractions to fail and shatter (Photos, 1989). This explains why almost all nickel-rich iron recorded within man-made steel artefacts is within the range of 3-5 wt. % nickel (Varoufakis, 1982).

Figure 7: Gibeon, a coarse octahedrite iron meteorite, individual piece and a cut slice, polished and etched in weak acid displaying the Widmanstätten structure, scale bar 2cm.

Historic working of iron meteorites

The earliest known examples of iron use are three balls believed to be tools found in a grave at Tepe Sialk, Iran dating to approximately 4000 BCE (Halm, 1939). Other examples of iron meteorites, more highly worked, have been discovered by other prehistoric cultures such as the Inuit of Cape York Greenland. They made use of three large iron meteorites at a location known by local people as 'iron mountain', and named the three meteorite masses the tent, the dog and the woman. These were used as an important source of iron where fragments were broken away from the large masses by use of hard stones which they then hammered with small boulders into small flat thin sheets to be used functionally, frequently as knife blades or harpoon tips (Buchwald, 1992). We also see evidence of the use of meteorites by Native American Indians at sites such as Hopewell mounds, Ohio, where a large number of objects were excavated from cremation burial mounds frequently as jewellery including beads and ear spools (Prufer, 1961; Grogan, 1948). These beads are of a very similar size and shape to the Gerzeh beads, and most likely were produced by people with a similar level of technology; other tomb contents being comparable from the two sites include canine teeth and exotic objects such as non-local shells and stones. There is also broader evidence of American Indians having an awareness of meteorites beyond irons, an example being the inclusion of a stony meteorite now known as Winona found within a burial site in its own small rock lined tomb wrapped within a shroud at the prehistoric native American village of Elden Pueblo, 5 miles north east of Winona, Arizona (Mason and Jarosewich, 1967).

Other meteorites have been discovered during archaeological excavation, but it is frequently difficult to determine how the meteorite was perceived and therefore the meaning of the meteorites presence. An example of this was encountered with the discovery at the British iron age hill fort at Danebury, where the stony meteorite now known as Danebury, was discovered at approximately 50cm depth in an unused grain storage pit at the approximate centre of the site (Cunliffe, 1984) (figure 8). It is yet to be determined if the site inhabitants placed the stone here for a special purpose. Meteorite iron has also frequently found to be employed within ceremonial goods such as in the case of the iron blades of broad and dagger axes from China, 1000 BCE (Gettens et al., 1971).

Figure 8: Optical image of the prehistoric Egyptian Gerzeh bead
(The Manchester Museum, 5303) with the British Iron Age Danebury meteorite
(Hampshire County Museums service), scale bar 1cm.

Conclusions

The interactions between people and meteorites are diverse, ranging from simple random small scale impacts witnessed by few people through to entire communities dependent on iron from large iron meteorite impacts. In some cases meteorites are valued simply as a useful source of iron, in others they are venerated such as the Kaaba stone at Mecca. The Gerzeh beads are one of the earliest known examples of humans interacting with meteorites, undoubtedly at some point in time ancient Egyptians would have developed knowledge of material falling from the sky but evidence to document this is scarce. Many potential links are known, such as references within the pyramid texts linking iron and the sky, later the creation of a new term for iron ($bi3$-n-pt 'iron from the sky') which was used from around the start of the 19th Dynasty (Bjorkman, 1973) strongly suggests knowledge of meteorite iron. Links have also been proposed between iron and the blades used in the opening of the mouth ceremony (Roth, 1992). The results reported here form the first scientific identification of metallic meteorite iron in ancient Egyptian artefacts. It supports theories of the cold working of iron meteorite fragments by ancient Egyptians into important funerary objects.

Acknowledgments

This work was supported by STFC and EPSRC. We thank K. Exell (formerly of The Manchester Museum, now at UCL Qatar), B. Sitch (The Manchester Museum) and C. Price (The Manchester Museum) who made this study possible by loan of a Gerzeh iron bead, A. Stevenson (Pitt Rivers Museum) for advice on the Gerzeh site, A. Tindle (Open University) for provision of image used in figure 3. Hampshire County Museums Service for access to the Danebury meteorite. The Petrie Museum of Egyptian Archaeology, UCL, for provision of figure 1, The Pitt Rivers Museum for provision of figure 2.

References

Bjorkman, J.K., 1973. Meteors and meteorites in the ancient near east. *Meteoritics and Planetary Science*, 8, pp.91-132.

Buchwald, V.F., 1992. On the use of iron by Eskimos in Greenland. *Materials Characterisation*, 29, p.2.

Burke, J.G., 1986. *Cosmic Debris: Meteorites in History*. Berkeley: University of California Press.

Cunliffe, B.W., 1984. *Danebury: an Iron Age hillfort in Hampshire*, Research Report 52. London: Council for British Archaeology.

Desch, C.H., 1928. *Report on the Metallurgical Examination of Specimens for the Sumerian Committee of the British Association.* Reports of the British Association for the Advancement of Science, Section H. Glasgow: The British Association.

El-Gayer, E.S., 1995. Predynastic iron beads from Gerzeh. *Institute for Archaeo-metallurgical Studies*, 19, pp.11-12.

Gettens, J.R., Clarke, R.S. and Chase, W.T., 1971. Two early Chinese bronze age weapons with meteoric iron blades. *Occasional Papers,* 4 (1). Freer Gallery of Art, Washington DC: Smithsonian Institution Press).

Golden, D.C., Ming, D.W. and Zolensky M.E., 1995. Chemistry and mineralogy of oxidation products on the surface of the Hoba nickel-iron meteorite. *Meteoritics*, 30, pp.418-422.

Grogan, R.M., 1948. Beads of meteoric iron from an Indian mound near Havana, Illinois. *American Antiquity*, 13 (4), pp.302-305.

Halm, L., 1939. Analyze chimique et etude micrographique dequelques pieces de metal et de ceramique provenant de Sialk. In: R. Ghirshman, ed. 1939. *Fouilles de Sialk, 2*. Paris: P. Geuthner. p.206.

Hutchinson, R., 2004. *Meteorites: A Petrologic, Chemical and Isotopic Synthesis*. Cambridge: Cambridge University Press. pp.322-338.

Mason, B. and Jarosewich, E., 1967. The Winona meteorite. *Geochimica et Cosmochimica Acta,* 31 (6), pp.1097-1099.

Petrie, W.M.F., 1886. *Naukratis I*. London: The Egypt Exploration Fund.

Petrie, W.M.F. and Wainwright, G.A., 1912. *The Labyrinth, Gerzeh and Mazghuneh*. London: British School of Archaeology in Egypt.

Petrie, W.M.F., 1939. *The making of Egypt*. London: Sheldon Press.

Photos, E., 1989. The question of meteoritic versus smelted nickel-rich iron: archaeological evidence and experimental results. *World Archaeology*, 20 (3), pp.403-421.

Piaskowski, J., 1982. A study of the origin of the ancient high nickel iron generally regarded as meteoritic. In: T.A. Wertime and S.F. Wertime, eds. 1982. *Early Technology*. Washington: Smithsonian Institution. pp.237-423.

Prufer, O.H., 1961. Prehistoric Hopewell Meteorite Collecting Context and Implications. *The Ohio Journal of Science* 61 (6), pp.341-352.

Richmond, A., 2005. The ethics checklist – ten years on. *V&A Conservation Journal* 50, pp.11–14.

Roth, A.M., 1992. The *psš-kf'* and the 'opening of the mouth' ceremony: a ritual of birth and rebirth. *Journal of Egyptian Archaeology*, 78, pp.113-147.

Varoufakis, G., 1982. The origin of Mycenaean and Geometric iron on the Greek mainland and in the Aegean islands. In: J.D. Muhly, R. Maddin and Karageorghis, V. eds. *Early Metallurgy in Cyprus*. Nicosia: Pierides Foundation. pp.315-322.

Wainwright, G.A., 1912. Pre-dynastic iron beads in Egypt. *Revue Archéologique*, 19, pp.255-259.

Wainwright, G.A., 1932. Iron in Egypt. *The Journal of Egyptian Archaeology*, 18 (1-2), pp.3-15.

Wasson, J.T., 1985. *Meteorites: Their Record of Early Solar System History*. New York: W. H. Freeman.

Imaging and analysis of skeletal morphology: New tools and techniques

Norman MacLeod

Palaeontology, The Natural History Museum, London, UK

Abstract

The ability to represent the geometric structure of specimens, bones and artifacts, and to analyse their positions, sizes and shapes, is revolutionising the illustration and scientific investigation of all aspects of vertebrate anatomy, including the study of skeletal pathology. Researchers have long had access to single scan X-rays and, more recently, digital photographs of specimens including detailed photomicrographs of pathological structures and surfaces. However, until recently the analysis of these materials has typically been undertaken via qualitative assessment, by eye, and as such relies heavily on the subjective experience and judgement of the observer. While this is satisfactory for some types of structures, some types of analyses, and some types of hypothesis tests, the inherently 3D character of many pathological structures represents a source of information that (now) can and should be assessed using data techniques and methods commensurate with the geometric character of these structures. Imaging and data collection devices such as microCT (X-ray) scanners, laser scanners (fixed, portable, and handheld), and optical scanners (designed for macroscopic and microscopic analyses) are now available for the collection of 2D and 3D (palaeo)pathological data. Similarly, geometric data-analysis methods for landmarks, outlines, and surfaces, in addition to various artificial intelligence, computer vision, and machine learning approaches, can facilitate the quantitative study of anatomical structures, both normal and pathological. These instruments and algorithms extend the range and detail of human senses in ways that demonstrably improve the speed, objectivity, reproducibility, and accuracy of anatomical assessments and so improve diagnoses and interpretations based thereon. Such technologies and procedures cannot, and will not, replace the need for human training, experience and judgement in any of these areas. Rather, the establishment of reciprocally illuminating partnerships with such technologies and procedures will free researchers, medical professionals, and technicians from the geometrically complex, but routine and repetitive, tasks that human visual and cognitive systems are not designed to deliver rapidly and/or with high levels accuracy, thereby allowing human specialists to focus on the conceptual, integrative, and adjudicative tasks for which they have no equal and from which genuine advances in our understanding of natural processes have always derived.

Introduction

The natural world exists in a geometric space. Any description of natural forms made on the basis of data extracted from images, scans or other representations of natural geometries and/or measurements that do not preserve the physical geometries of natural objects should be regarded (and labelled) as results derived from 'apparent' geometry that illustrate or summarise 'apparent' patterns of similarity and difference. Yet, for literally hundreds of years science in general, and anthropology in particular, has relied on precisely these sorts of measurements to test morphology-based hypotheses.

Why did this situation persist for such a long time? How have recent conceptual and technological advances improved matters? What tools, data, techniques and approaches should medical anthropologists be using to support the illustration and quantitative analysis of morphological data? And what research is being conducted at the moment that will improve this situation still further in the future?

The classical biometric approach to morphological analysis

The classic approach to quantitative morphological analysis (also called biometrics) involved the description of morphology via the collection of a set of distances recording the separation of geometric points located on the object that could be identified on every specimen within the sample of interest. There are many systems that have been employed for describing patterns of morphological variation using distance measurements, some that have become more-or-less standardised for certain parts of certain groups (e.g., human crania, see figure 1). Others have been used in a more idiosyncratic manner for particular investigations.

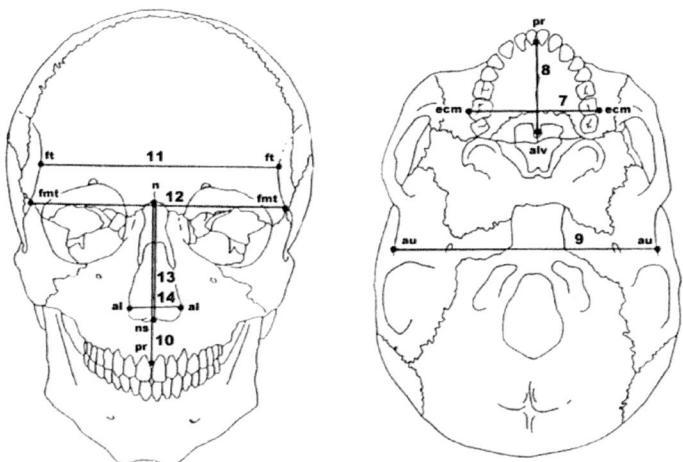

Figure 1: Standard human cranial landmarks and measurements. Landmarks: alare (al), alveolon (alv), bregma (b), basion (ba), dacryon (d), ectoconchion (ec), ectomolare (ecm), euryon (eu), frontomalare temporale (fmt), frontotemporale (ft), gnathion (gn), gonion (go), infradentale (id), nasion (n), nasospinale (ns), opisthion (o), opisthocranion (op), prosthion (pr), zygion (zy). Measurements: (7) maxillo-alveolar breadth [ecm-ecm], (8) maxillo-alveolar length [pr-alv], (9) biauricular breadth [au-au], (10) upper facial height [n-pr], (11) minimum frontal breadth [ft-ft], (12) upper facial breadth [fmt-fmt], (13) nasal height [n-ns], (14) nasal breadth [al-al]. From Buikstra and Ubelaker, 1994 and Moore-Jansen et al., 1994.

Most commonly these measurements were collected with physical callipers that assessed the separation between landmark points in the three dimensions of descriptive geometry. However, far from forming a complete description even of the structures measured (e.g., the nareal opening or cranial dental arcade shown in figure 1), these measurements only recorded aspects of these structures such as total length or maximal width. This problem was compounded by the fact that, because distances are simply scalar magnitudes in which no relative geometric information resides, it was impossibly laborious to reconstruct any but the most trivial aspects of the actual geometry from the distance measurements so collected. Moreover, even if the supplementary information required to attempt such reconstruction was collected, this information was rarely, if ever, employed in the morphological analysis, especially in time prior to the advent of digital computers.

This is not to say that the collection of distance-based information in the classical period of biometric analysis was pointless. Indeed, such information still serves as the standard for human craniometric analysis (Howells, 1973; Buikstra and Ubelaker, 1994; Moore-Jansen et al., 1994). Standard anthropological and forensic morphometric identification packages such as CRANID and FOREDISC (http://osteoware.si.edu/guide/craniometrics and https://web.dii.utk.edu/FORDISC_V2/Login.aspx respectively) continue to be based, in part, on such data. Rather, my purpose is to raise the question of whether these types of variables and this approach to morphological description are optimal for the purpose of characterising or analysing biological forms in either generalised geometric or specific medical contexts.

Multivariate vs superposition-based morphometrics

Following the appearance of digital computers on university campuses in the 1950s, the analysis of large sets of distance measurements became tractable. To a large extent, these data were still collected in the same way they have always been collected, with either straight rulers used on two-dimensional (2D) images of orientated specimens or from X-radiographs or with physical (3D) callipers (figure 2A). In terms of distance-based descriptions of morphological variation, this transition was straight-forward, involving little more than the assembly of large suites of distance-based variables into data matrices that were submitted to cluster analysis, principal components analysis (PCA), canonical variates analysis (CVA), etc. (Blackith and Reyment, 1971; Pimentel, 1979). This school of biometric analysis came to be known as multivariate morphometrics. However, with the advent of electronic digitising tablets it also became possible — though still laborious — to collect geometric information about boundary outline curves as well as landmarks (figure 2B). This was done by manually digitising a series of landmark points along a curve of interest and then (usually) using interpolation methods to estimate a set of equiangular or equally spaced boundary outline landmarks that collectively represented the form of interest. Because the location of any landmark point used to describe these boundary outline curves was dependent on the location of the preceding boundary landmark point in the series, their specificity and independence of location relative to underlying biological structures was regarded as not being as great as true landmarks that were used to locate unified biological structures relative to each other. Accordingly, the members of these boundary outline landmark sets have come to be termed semilandmarks in the contemporary morphometric literature (see Bookstein, 1990a; 1991).

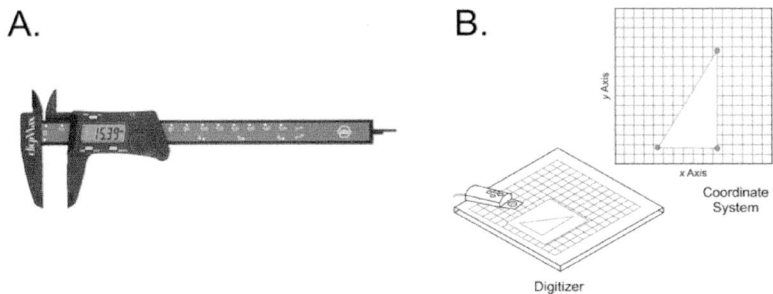

Figure 2: A. Digital callipers, used for measuring 3D distances between landmark points from specimens. B. Schematic drawing of a 2D digitising tablet; used for recording the positions of landmarks and semilandmarks from 2D drawings, photographs, or scans. These were the primary geometric data collecting devices prior to the advent of 3D electronic scanners (see figure 4).

The ability of manual digitisers to collect sets of semilandmark data that collectively describe biological curves of interest provided an incentive to develop new methods of geometric data analysis to solve the particular problems such data posed. At first this analysis drew heavily on two mathematical methods both of which focused on correcting the semilandmark data for various nuisance factors, but were not part of the comparisons biologists wished to make. These nuisance factors were differences in position within the coordinate system, differences in rotation and (in many, but not all cases) differences in scale. The primary biometric uses of both Fourier analysis (Christopher and Waters, 1974; Younker and Ehrlich, 1977) and Procrustes analysis (Benson et al., 1982; Chapman, 1990; Rohlf, 1990) were to redescribe semilandmark-quantified shapes in such a way as to normalise semilandmark data for the geometric effects of these factors. Once this normalisation had been achieved the resulting redescriptions of shape variation in terms of shape coordinates could then be submitted to any of a number of standard multivariate data analysis procedures to summarise patterns of variation, test for covariation of shape variation with external factors, maximally separate a priori designated groups, etc. The school of morphometrics that used Fourier or Procrustes methods to align landmark and boundary outline semilandmark data prior to analysis was termed 'superposition morphometrics'.

Throughout the late 1970s and early 1980s adherents to both multivariate and superposition schools pursued their analyses independently with some practitioners on both sides adopting a position of mutual antagonism with regard to the other (e.g., see Bookstein et al., 1982 and Ehrlich et al., 1983). The multivariate morphometric school felt the superposition school had sacrificed biological interpretability for broadly specified completeness of coverage and vice versa. In retrospect though, what the hardening of these philosophical positions did was set the stage for a unification of both schools and, even more importantly, a fundamental reconceptualisation of what form (= size and shape) represents from a mathematical point of view.

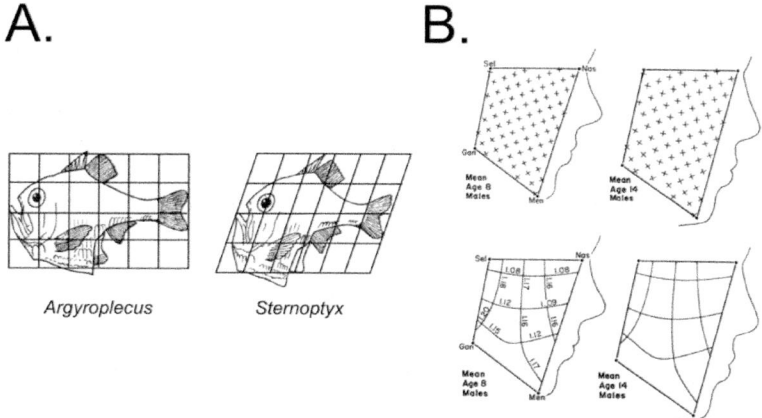

Figure 3: A. d'Arcy Thompson's drawing of a transformation grid depicting the overall character of the geometric transformation implied by the comparison of two fish morphologies. Thompson's analysis demonstrated that what can appear to be morphologically complex changes can have a simple underlying geometry that will only become evident upon close tracking of relative landmark displacements. From Thompson, 1917. B. Bookstein's original attempt to quantitise Thompson's conceptual transformation grid approach as a series of biorthogonal axes calculated from regional deformation ellipses that express degrees and directions of localised relative shape change. Nas = Nasion, Men = Menton, Gon = Gonion, Sel = Sella. From Bookstein, 1986.

The morphometric synthesis

Among those who followed the school of multivariate morphometrics Fred Bookstein was unusual in having developed an early interest in the characterisation of shape change as a deformation in the manner most cogently expressed up to that point by d'Arcy Thompson's transformation grid concept (Thompson, 1917; 1942; figure 3A) and in having acquired the mathematical skills to turn his conceptual desires into computational reality. Bookstein's biorthogonal grid method (Bookstein, 1978; Bookstein et al., 1985) was an early attempt to place Thompson's concept on a solid mathematical footing and arguably the most successful such attempt to that time (figure 3B). Following a very innovative attempt to incorporate the information necessary to reconstruct geometries from multivariate morphometric distance data (Strauss and Bookstein, 1982), Bookstein turned his attention to landmark coordinate data as a source of information about biological geometries that preserves the geometry of the sampled form. Bookstein's key insight was the reconceptualisation of shape analysis not as a collection of geometrically disembodied scalar magnitudes, but as a geometrically unified set of landmark configurations. This shift of the frame of biometric reference from the distance variable to a constellation of landmark points in 2D (and later 3D) space was always present in the ordination plots of PCA scores derived from multivariate morphometric data. But until Bookstein's work on the unification of multivariate and superposition morphometrics (Bookstein, 1990a; 1990b; 1991) the implications of the plots morphometricians had been constructing for decades had remained obscure.

This is not the place to describe the morphometric synthesis in detail. Readers interested in this topic should consult Bookstein (1993) for an insider's history, Bookstein (1991) for mathematical treatments, and/or MacLeod (2009) for a non-mathematical description. Key aspects and references to this fascinating conceptual revolution are listed below.

- Description of an F-ratio test for shape difference (Goodall, 1983)
- First description of shape coordinates and proposal of a T2 statistical test for shape difference (Bookstein, 1984)
- Synthesis of the Goodall and Bookstein tests and demonstration that they operated in a tangent space to the landmark-defined shape manifold (Kendall, 1984)
- First description of Bookstein shape coordinates (Bookstein, 1986)
- First description of the thin plate spline and demonstration that it defined a highly useful feature space for shape analysis problems (Bookstein, 1989)
- First comprehensive description of geometric morphometrics (Bookstein, 1990a; 1990b)
- Canonical description of geometric morphometrics (Bookstein, 1991)

Many books of collected essays, single-authored book-length treatments, review papers, methods papers, and (especially) applications papers have followed in the intervening two decades (see Adams et al., 2004; Adams et al., 2013). Until quite recently though, geometric morphometrics was felt to pertain only to collections of landmark coordinates and has been applied in the context of only 2D images of 3D objects.

Landmarks vs outlines

Owing to concerns over the manner in which the similarity of two shapes is dependent on the amount and manner in which data are collected from them, and the range of approaches that might be used to specify semilandmarks along a form, Bookstein (1990a; 1990b; 1991) originally rejected the idea that boundary outline semilandmarks could play any useful role in geometric morphometric analysis. However, (1) the demand for quantitative analyses of biological structures on which landmarks were difficult to locate in sufficient numbers to make the results useful, (2) the demonstration that, past a certain point all approaches to semilandmark sampling converge in the same geometric result (see MacLeod, 1999), (3) the heuristic advantages of modelling the results of semilandmark analyses to aid in the assessment and interpretation of specimen ordinations in geometric spaces (see MacLeod, 1999), and (4) the practical advantages of being able to combine information from landmarks internal to the form outline and outline semilandmarks in the same analysis (see Figueirido et al., 2011) has driven increasing interest in the use of semilandmarks in morphometric analyses. Techniques for implementing such analyses include Fourier analysis (incl. edgels, Bookstein and Green, 1993; Bookstein, 1997; radial Fourier, see MacLeod, 2011a; Z-R Fourier, MacLeod, 2011b; elliptical Fourier, Ferson et al., 1985; MacLeod, 2012a) and eigenshape analysis (Lohmann, 1983; MacLeod, 2012b; Bookstein and Ward, 2013; including extended eigenshape analysis, MacLeod, 2012b). Of these only eigenshape and extended eigenshape support the full range of flexibility with regard to including different types of data in the same analysis.

Geometric surfaces and the 3D revolution

Despite the impressive range of analyses and scope of applications the geometric approach to morphometrics was capable of supporting by 2000, and despite the fact that its methods had, by this time, been extended to the consideration of both 3D landmark coordinate and 3D semilandmark coordinates, the analysis of true 3D surfaces remained an unmet challenge. To some extent this was the result of practical constraints, the chief one being that 3D digitisers were quite expensive (and so quite rare on university campuses) to that point. Without the means to collect data from 3D surfaces there was no particular need to consider the question of how geometric morphometrics could be extended to operate on such data. This began to change shortly after 2000, though, with the appearance of relatively inexpensive mechanical, laser, and optical 3D digitisers to meet the needs of the computer-aided design and computer gaming industries (figure 4). Once the tools for collecting such data were placed in the hands of morphometricians, they were quick to respond with new algorithms and procedures to take advantage of these data.

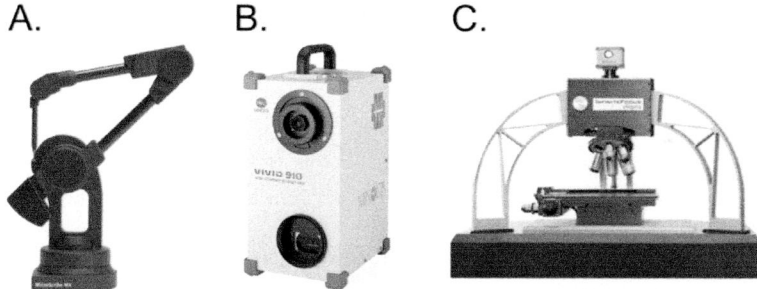

Figure 4: Three different types of modern 3D digitisers.
A. The MicroScribe, a physical digitiser that collects 3D coordinate data one point at a time via placement of the tip of the stylus on the landmarks point and activating a foot switch to record the tip's position. B. A Konica-Minolta laser scanner that collects a cloud of 3D coordinate point data via measurement of reflections of light as a laser beam scans across an object. C. An Alicona Infinite-Focus Microscope that reconstructs the 3D geometry of microscopic objects via reflected light using a stereoscopic procedure.

An early demonstration of the principles of 3D surface construction and modelling was provided in a study of hominin crania by Gunz et al. (2005, see also Gunz et al., 2009). Unfortunately, neither of these articles, nor a subsequent description (Gunz and Mitteroecker, 2013), provide details for the most critical part of this analysis, the interpolation of topologically homologous semilandmark points across the surface. Gunz et al. (2009) stated that this was accomplished using a complex multivariate regression procedure that used complete specimens of modern human crania as the basis for interpolation of a set of equivalent 3D semilandmark points (incl. estimation of the positions of points on missing sections of the crania) on a study set of ancient human crania. Obviously this procedure is not generalised and would require development of similar species-specific or group-specific regression models for locating semilandmarks on each category of objects selected for investigation. Nevertheless, a genuinely generalised approach to the specification of 3D surface semilandmarks has now been developed (MacLeod, 2008; see also Polly, 2008; Polly and MacLeod, 2008). This approach is capable of providing accurate semilandmark data for the analysis of any 3D surface whose outline is single valued with regard to a sampling-control chord drawn connecting relocatable landmarks on the surface's outline and midline (see figure 5). To date this approach has been used to analyse 3D surface variation in bivalve shells, vertebrate post-cranial skeleton anatomy, vertebrate crania and mandibles, vertebrate teeth, and archaeological artefacts.

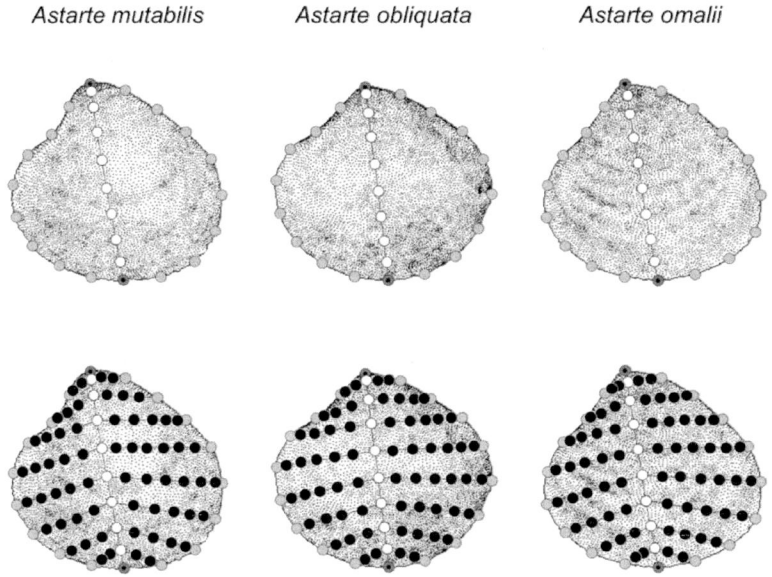

Figure 5: Steps in sampling surface morphology using the eigensurface procedure as applied to the three bivalve specimens. The dark grey symbols mark positions of the landmarks used to orient the semilandmark sampling grid, light grey symbols the outline semilandmarks, white the midline semilandmarks, and black the 'rib' semilandmarks that join the midline to the outline segments. These are 10-grids in which each half outline and the midline is represented by 10 semilandmarks. Spatial resolution of the sampling network is controlled by increasing or decreasing the number of equally spaced semilandmark points used to represent these dimensions of specimen form. See MacLeod (2008) for further discussion.

Computer vision and automated identification

To complete the circle we began in the Introduction, just as 1D linear distances between landmarks were used to represent the relative positions of 2D or 3D landmarks, and just as 2D and 3D landmarks and boundary outline coordinates are used to represent some of the characteristics of 3D surfaces, 3D surfaces are now being used to represent the forms of complex objects made up of multiple 3D surfaces. At the moment the collection and processing of the data used to represent 3D surfaces is complicated and both labour and computation-time intensive; often requiring the writing of specialist software. This effort is justified because 3D surfaces — especially when combined with 3D landmark data — represent the most complete characterisation that can be achieved of the forms of the specimens and objects biologists and anthropologists wish to study. But is it always necessary to go to the trouble of taking any measurements in order to compare the morphologies of a sample of objects or specimens? After all, the forms of these objects have already been sampled (at least in a sense) by the digital camera or scanner we used to capture the image of these objects in the first place.

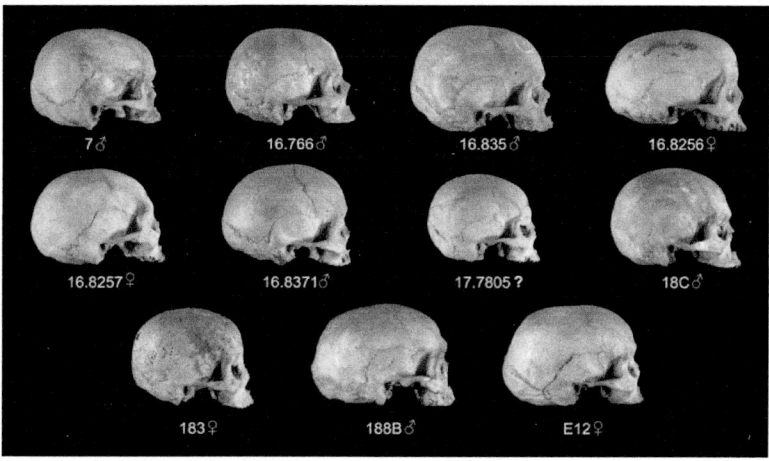

Table 1: A small sample of mixed male and female crania
from the Archaeological Survey of Nubia Online Database
(http://www.knhcentre.manchester.ac.uk/research/nubiaproject/)

A digital image of an object records a set of observations made on the object in precisely the same way linear distances, 2D or 3D landmarks, and 2D or 3D semilandmarks do. In the case of digital images the coordinate position of the object itself and parts thereof is fixed by the image frame. The observations then are the grey level brightness or colour values assigned to each picture element or pixel. Together, these values define the image frame and, in so doing, the form of the image (figure 6). Note that while this representation of the object is, strictly speaking 2D, systematic patterns of shading and/or colour variation across the pixel frame come together to provide signals as to the 3D form of the object or set of objects (table 1). This representation of the complex 3D geometries of many objects is far from perfect. But in most contexts it contains sufficient, and sufficiently well organised, information that those who look at the image are able to understand and appreciate its 3D form.

Once we've shifted our frame of reference and begun to see digital images as collections of morphometric data, it's only a short step to take to devise new, simple, flexible, and fast ways of analysing similarities and differences among a collection of objects. Take, for example, a collection of images of crania from the Archaeological Survey of Nubia (table 1). These images can be used in a morphometric analysis directly by downsampling the pixel grid to remove redundancy, converting the colour values to grayscale brightness values to minimise the effect of colour variation, and reformatting these brightness values into a column vector. Size differences among the crania can be either retained

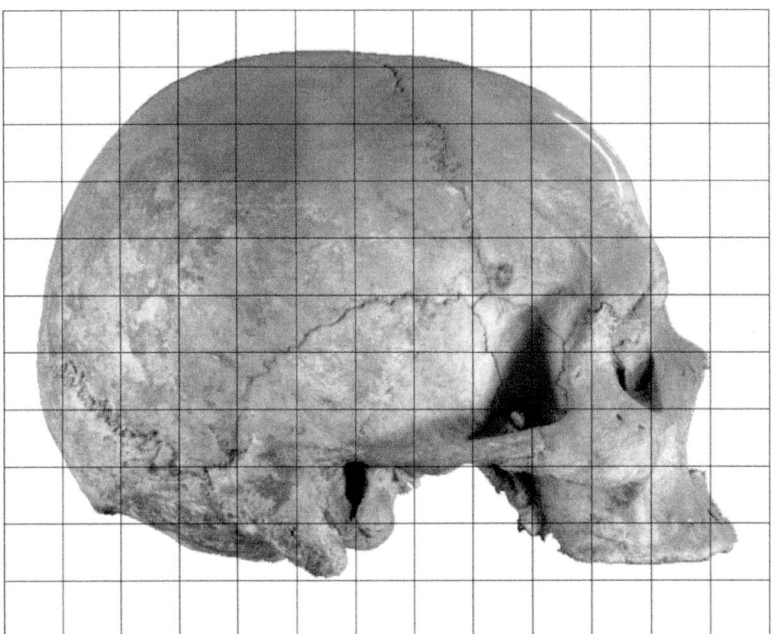

Figure 6: Theory behind use of pixel coordinate brightness/colour values as an indicator of 3D morphology. A digital image represents a rectilinear sampling surface that can be used to collect morphological data from any object as a pattern of brightness or colour values. Normally this pattern is reproduced in software to produce an image of the object of interest. However, the brightness/colour values themselves also represent data that has been collected systematically across the entire image. These data quantify the object's size, shape, outline, positions of component parts, surface markings, and details of its' surface texture. This redescription of the object's form can be compared directly with similar digitisations of the forms of other objects to represent a generalised measure of form similarity/difference between all pairs of objects in a sample. See below for an example.

or dispensed depending on the aims of the investigation during the downsampling procedure. For a 40 x 60 pixel grid, size-standardised representation of the information contained in these images, shape similarities and differences of this set of crania can be summarised on the first three principal component axes (figure 7).

As can be seen on these plots, shape variation is highly structured in this sample and appears to reflect sexual dimorphism. Male crania exhibit a much more spherical form in lateral view than female crania which are decidedly elongate. The juvenile cranium in the sample is quite distinct in shape from any of the adult crania, exhibiting an elongation reminiscent of an adult female. However, one of the female crania (183) plots well away from the main adult female point cloud in both plots. This cranium has the spherical character more strongly reminiscent of an adult male and the size of a juvenile (see table 1). Based on these results it seems likely this cranium has been mislabelled. Note also that the plot in figure 7B also indicates there are two distinct male morphs present in

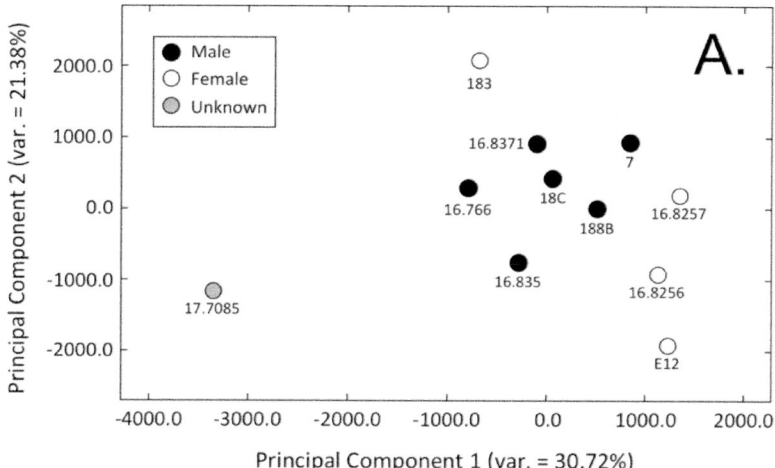

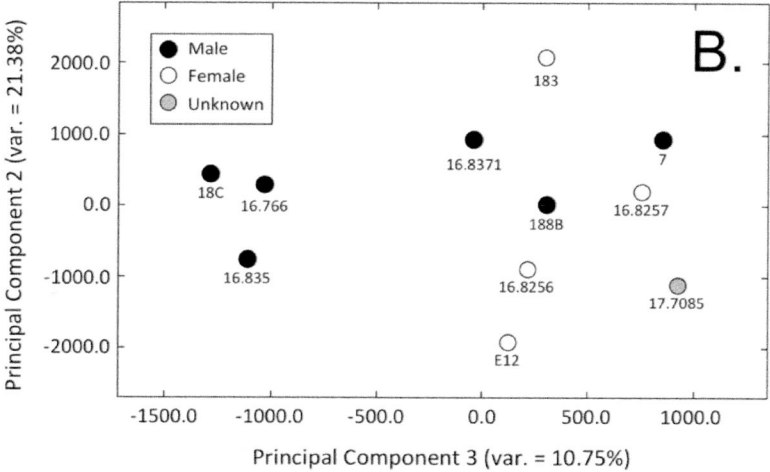

Figure 7: Principal component ordination of the specimen images shown in Table 1. Prior to analysis the images were downsampled to a 60 x 40 pixel grid, size-standardised, converted to greyscale and image-histogram normalised in order to reduce the effect of colour variations on the similarity assessments. The first three PCA axes represent the major directions of 3D shape variation in the sample. See text for discussion.

this sample, one characterised by dorso-ventral vaulting (18C, 16,766, 16.835) and the other by dorso-ventral flattening (16.8271, 188B, 7).

The ability to summarise the major modes of form variation in a sample of specimens quickly and accurately in the manner demonstrated above from the information contained in the images alone represents a new, exciting, and eminently practical tool morphologists in general, and archaeologists in particular, can exploit in the context of their scientific investigations.

Summary and prospectus

Currently the study of morphology is undergoing a renaissance in both theory and practice. Owing primarily to the ease and convenience with which this digital data can be collected, manipulated, and displayed, natural phenomena that were once apprehended and explored using non-visual methods (e.g., sounds, textures, tastes, and smells) or that could only be displayed visually as the result of a long and complex sampling procedure, are portrayed as visual forms, patterns, and objects routinely. While it is no coincidence that the unanticipated abundance of these visual data have spurred the development of new approaches to visual data analysis, many of the new tools that have been developed have yet to be applied over the full range of analytic contexts for which they are suited (see MacLeod, 2013 for an intriguing example).

Use of tools such as those reviewed above is required by all researchers who need to characterise, describe, compare, and identify objects based on their morphology. In particular the study of (palaeo)pathology has much to gain from the wider application of these tools and from moving to a quantitative approach to the study of all skeletal structures, both normal and pathological. Far from replacing the skills of morphological specialists, forming productive partnerships with researchers and technologists skilled in the 2D and 3D characterisation of form will improve the speed, consistency, accuracy, and scope of pathological analyses, eventually freeing pathologists to focus on the conceptual, integrative, and adjudicative tasks that no machine or algorithm can replicate. As always, the primary challenge this fortuitous situation entails is for researchers, students, databasers, geometers, and technologists to come together and combine their talents so that these new tools — and tools as yet undreamt of — can be applied to these new data to achieve our collective goal which is to understand the Earth's, and humanity's, past and present so that we might be in a better position to understand and make reasonable choices about its future.

Acknowledgements

I thank the editors of this volume for the invitation to contribute this manuscript to the volume.

References

Adam, D.C., Rohlf, F.J. and Slice, D.E., 2004. Geometric morphometrics: ten years of progress following the 'revolution'. *Italian Journal of Zoology*, 71, pp.5-16.
Adam, D.C., Rohlf, F.J. and Slice, D.E., 2013. A field comes of age: geometric morphometrics in the 21st century. *Hystrix*, 24, pp.13-20.
Benson, R.H., Chapman, R.E. and Siegel, A.F., 1982. On the measurement of morphology and its change. *Paleobiology*, 8, pp.328–339.

Blackith, R.E. and Reyment, R.A., 1971. *Multivariate morphometrics*. London: Academic Press.

Bookstein, F.L., 1984. A statistical method for biological shape comparison. *Journal of Theoretical Morphology*, 107, pp.475–520.

Bookstein, F.L., Chernoff, B., Elder, R., Humphries, J., Smith, G. and Strauss, R., 1985. *Morphometrics in evolutionary biology: the geometry of size and shape change, with examples from fishes*. Philadelphia: Academy of Natural Sciences of Philadelphia.

Bookstein, F.L. and Ward, P.D., 2013. A modified Procrustes analysis for bilaterally symmetrical outlines, with an application to microevolution in Baculites. *Paleobiology*, 39, pp.214–234.

Bookstein, F.L., 1978. *The measurement of biological shape and shape change*. Berlin: Springer.

Bookstein, F.L., 1986. Size and shape spaces for landmark data in two dimensions. *Statistical Science*, 1, pp.181–242.

Bookstein, F.L., 1989. Principal warps: thin-plate splines and the decomposition of deformations. *IEEE Transactions on Pattern Analysis and Machine Intelligence*, 11, pp.567–585.

Bookstein, F.L., 1990a. Analytic Methods: Introduction and Overview. In: F.J Rohlf and F.L. Bookstein, eds. 1990. *Proceedings of the Michigan Morphometrics Workshop*. Ann Arbor, MI: The University of Michigan Museum of Zoology, Special Publication 2.

Bookstein, F.L., 1990b. Higher-order features of shape change for landmark data. In: F.J Rohlf and F.L. Bookstein, eds. 1990. *Proceedings of the Michigan Morphometrics Workshop*. Ann Arbor, MI: The University of Michigan Museum of Zoology, Special Publication 2. pp.237-250.

Bookstein, F.L., 1991. *Morphometric tools for landmark data: geometry and biology*. Cambridge: Cambridge University Press.

Bookstein, F.L., 1993. A brief history of the morphometric synthesis. In: L.F. Marcus, E. Bello and A. García-Valdecasas, eds. 1993. *Contributions to Morphometrics*. Madrid: Museo Nacional de Ciendcias Naturales 8.

Bookstein, F.L., 1997. Landmark methods for forms without landmarks: Localizing group differences in outline shape. *Medical Image Analysis*, 1, pp.225–243.

Bookstein, F.L. and Green, W.D.K., 1993. A feature space for edgels in images with landmarks. *Journal of Mathematical Imaging and Vision*, 3, pp.213–261.

Bookstein, F.L., Strauss, R.E., Humphries, J.M., Chernoff, B., Elder, R.L. and Smith, G.R., 1982. A comment on the uses of Fourier methods in systematics. *Systematic Zoology*, 31, pp.85–92.

Buikstra, J.E. and Ubelaker, D.H., 1994. *Standards for data collection from the human skeletal remains*. Fayetteville, Arkansas: Arkansas Archaeological Survey.

Chapman, R.E. 1990. Conventional Procrustes approaches. In: F.J Rohlf and F.L. Bookstein, eds. 1990. *Proceedings of the Michigan Morphometrics Workshop*. Ann Arbor, MI: The University of Michigan Museum of Zoology, Special Publication 2.

Christopher, R.A. and Waters, J.A., 1974. Fourier analysis as a quantitative descriptor of miosphere shape. *Journal of Paleontology*, 48, pp.697–709.

Ehrlich, R., Pharr, R.B. and Healy-Williams, N., 1983. Comments on the validity of Fourier descriptors in systematics: a reply to Bookstein et al. *Systematic Zoology*, 32, pp.202–206.

Ferson, S., Rohlf, F.J. and Koehn, R.K., 1985. Measuring shape variation of two-dimensional outlines. *Systematic Zoology*, 34, pp.59–68.

Figueirido, B., MacLeod, N., Krieger, J., De Renzi, M., Pérez-Claros, J.A. and Palmqvist, P., 2011. Constraint and adaptation in the evolution of carnivoran skull shape. *Paleobiology*, 37, pp.490–518.

Goodall, C. R. 1983. *The statistical analysis of growth in two dimensions*. PhD Dissertation, Department of Statistics, Harvard University, Cambridge, Massachusetts.

Gunz, P. and Mitteroecker, P., 2013. Semilandmarks: a method for quantifying curves and surfaces. *Hystrix*, 24, pp.109-115.

Gunz, P., Mitteroecker, P. and Bookstein, F.L., 2005. Semilandmarks in three dimensions. In: D.E. Slice, ed. *Modern Morphometrics in Physical Anthropology*. New York: Kluwer Academic/Plenum Publishers.

Gunz, P., Mitteroecker, P., Neubauer, S., Weber, G.W. and Bookstein, F.L., 2009. Principles for the virtual reconstruction of hominin crania. *Journal of Human Evolution*, 57, pp.48–62.

Howells, W.W., 1973. *Cranial variation in man: a study by multivariate analysis of patterns of difference among recent human populations*. Harvard University: Cambridge, Massachusetts.

Kendall, D.G., 1984. Shape manifolds, procrustean metrics and complex projective spaces. *Bulletin of the London Mathematical Society*, 16, pp.81–121.

Lohmann, G.P. 1983. Eigenshape analysis of microfossils: A general morphometric method for describing changes in shape. *Mathematical Geology*, 15, pp.659-672.

MacLeod, N., 1999. Generalizing and extending the eigenshape method of shape visualization and analysis. *Paleobiology*, 25, pp.107–138.

MacLeod, N., 2008. Understanding morphology in systematic contexts: 3D specimen ordination and 3D specimen recognition. In: Q. Wheeler, ed. *The New Taxonomy*. London: CRC Press, Taylor and Francis Group.

MacLeod, N., 2009. Shape theory. *Palaeontological Association Newsletter*, 71, pp.34–47.

MacLeod, N., 2011a. Semilandmarks and radial Fourier analysis. *Palaeontological Association Newsletter*, 76, pp.25-42.

MacLeod, N., 2011b. The center cannot hold I: Z-R Fourier analysis. *Palaeontological Association Newsletter*, 78, pp.35–45.

MacLeod, N., 2012a. The center cannot hold II: elliptic Fourier analysis. *Palaeontological Association Newsletter*, 79, pp.29–42.

MacLeod, N., 2012b. Going round the bend: eigenshape analysis I. *Palaeontological Association Newsletter*, 80, pp.32–48.

MacLeod, N., 2012c. Going round the bend II: extended eigenshape analysis. *Palaeontological Association Newsletter,* 81, pp.23–39.

MacLeod, N., 2013. Geometric morphometric approaches to acoustic signal analysis in mammalian biology. In: H. Iwata, H. Tatsuta, Y. Yoshioka, S. Ninomiya and P.E. Lestrel, eds. 2013. *3rd International Symposium of Biological Shape Analysis*, 2013. Tokyo, Japan: University of Tokyo, 16.

Moore-Jansen, P.M., Ousley, S.D. and Jantz, R.J., 1994. *Data collection procedures for forensic skeletal material*. Knoxville, Tennessee: University of Tennessee.

Pimentel, R.A., 1979. *Morphometrics: the multivariate analysis of biological data*. Dubuque, IA: Kendall Hunt.

Polly, P.D. 2008. Adaptive zones and the pinniped ankle: a 3D quantitative analysis of carnivoran tarsal evolution. In: E. Sargis and M. Dagosto, eds. 2008. *Mammalian Evolutionary Morphology: A Tribute to Frederick S. Szalay*. Dordrecht, The Netherlands: Springer.

Polly, P.D. and MacLeod, N., 2008. Locomotion in fossil Carnivora: an application of the eigensurface method for morphometric analysis of 3D surfaces. *Palaeontologia Electronica*, 11, p.13.

Rohlf, F.J., 1990. Rotational fit (Procrustes) methods. In: F.J. Rohlf and F.L. Bookstein, eds. 1990. *Proceedings of the Michigan Morphometrics Workshop*. Ann Arbor, MI: The University of Michigan Museum of Zoology, Special Publication 2.

Strauss, R.E. and Bookstein, F.L., 1982. The truss: body form reconstruction in morphometrics. *Systematic Zoology*, 31, pp.113–135.

Thompson, D.W., 1917. *On growth and form*. Cambridge: Cambridge University Press.

Thompson, D.W., 1942. *On growth and form*. Cambridge: Cambridge University Press.

Younker, J.L. and Ehrlich, R., 1977. Fourier biometrics: harmonic amplitudes as multivariate shape descriptors. *Systematic Zoology*, 26, pp.336-342.

Mummies on rails

Ahmad Alam,[1] Ian Dunlop,[1] Robert Stevens,[1] Andrew Brass,[1] Jenefer Cockitt,[2] Rosalie David[2] and Ryan Metcalfe[2]

[1]School of Computer Science, The University of Manchester, Manchester, UK
[2]KNH Centre for Biomedical Egyptology, The University of Manchester, Manchester, UK

Abstract

Ancient Egyptian mummies are an important source of archaeological, historical and scientific data, which is varied and heterogeneous - ranging from Edwardian archaeological dig records to that obtained from more recent autopsies and medical scanning. Collating, linking and sharing this data presents considerable challenges, particularly as so much of the data is not readily accessible in a computationally readable form. Electronic recording has to date been limited to spreadsheets, isolated databases and standalone applications. This paper details the development of a system for comprehensively recording data related to mummies, with the goal of promoting open data and data sharing in an accessible manner. The system, the Mummy Electronic Patient Record (MEPR) has been implemented as a web application, in the Ruby on Rails development framework, which can be accessed globally by authorised users, and populates a scalable relational database. A particular advantage of this infrastructure is that it can be run entirely on the web and without the need for users to install any applications or databases. MEPR has been used to capture and present the data from the Dakhleh Oasis Mummies described by A. C. Aufderheide and colleagues.

Introduction: objectives of the research

The aim of this work is to produce an open data framework for archaeologists. Ancient Egyptian mummies have been chosen as the initial case study. This will then be expanded into other areas if it is successfully adopted by its target community. Electronic records will provide an important tool for the scientific research being carried out on diseases in antiquity, the geographical dispersal and movement patterns of such diseases through the course of history, and other related research work by facilitating access to large data sets. Although there is electronic recording of archaeological data, it is disparate and typically recorded by scientists working in their own particular field, using basic desktop applications.

Archaeology overlaps with most scientific disciplines such as biology, geology and chemistry. Although cross-disciplinary partnerships between scientists in these fields and archaeologists are frequently reported at conferences and in the literature, the storage and dissemination of data outside these routes is not common.

The collection of data, in a format that will stand the test of time, has become more urgent due to the increasing number of repatriations of archaeological artefacts to their countries of origin (Stodder, 2012). Additionally ancient human remains are a fragile and finite resource, and repeated examinations cause them to degrade.

The Dakhleh Oasis Mummy Project

The Dakhleh Oasis Mummy Project is a significant subset of the expansive Dakhleh Oasis project that was formally initiated in 1978 (Mills, 2013) and is still active, with archaeologists from at least nine institutions working on it at the time of writing. Part of the project was an extremely detailed examination of 46 mummified remains conducted by AC Aufderheide and colleagues in 1993 and 1998 (Aufderheide, 2003). A 13 page form with textual descriptions, diagrams and tick boxes etc. was developed by the University of Minnesota to capture the information required. As an autopsy is by definition destructive, the data obtained represents a unique dataset that cannot be replicated. The need for autopsying has largely been obviated by the development of non-destructive techniques e.g. CT scans.

From an information systems perspective, the dataset is unusual in that it is a very rich (detailed) but small dataset, i.e. the number of fields (>300) exceeds the number of records (46) by a substantial margin. In contrast a national address database would typically have a score fields, but millions of records.

Significance of the Dakhleh Oasis Project and its data

The Dakhleh Oasis Project's continuing significance and importance is supported by the number of its institutions participating, and the significant number of archaeological studies that have arisen. Examples of these are wide and varied, as demonstrated by the 2012 Seventh International Conference of the Dakhleh Oasis project (Anon., 2012a) that had 70 papers presented. The research topics covered a broad range of time periods, from the Pharaonic to the Islamic period. The mummies have also been the subject of a large amount of scientific analysis (e.g. Aufderheide et al., 2003).

Present state of the data, and how it is used

There appears to be no formal sharing of information that has been gathered, or of new information generated as a result of research, as there is no central Dakhleh Oasis Project Website. Each institution involved hosts its own work in the project. The Dakhleh Trust a registered charity that funds such research appears to have no web presence.

Despite consulting authoritative works such as Scientific Study of Mummies (Aufderheide, 2003) and The Cambridge Encyclopaedia of Human Paleopathology (Aufderheide and Rodriguez-Martin, 1998), no standard form for recording data related to mummies could be found, some minor tables of data were identified, but nothing substantive. Although copies of the autopsy records were available, these had been sent as plain paper faxes direct to the KNH Centre for Biomedical Egyptology by Prof. Aufderheide as additional information pertinent to a donation of tissue samples from the autopsies. Only a single copy was available at the start of this work, with sharing restricted to direct contact. To perform any analysis, the forms have to be physically searched, one by one, for the correlations that are being sought. This is inherently an arduous task, and only the results obtained for any such analysis are likely to be passed on, rather than each step taken.

Hence anyone else analysing the data may well end up repeating much of the effort of a previous analysis.

Benefits an information system can offer

Moving the paper records into any electronic format would have an immediate benefit. Simply scanning them as images would be the first step in improving data sharing. The scanned collection of records could then be placed on the Web as a shared file. However this does not help with how the forms are used. Simply substituting a paper sheet for a computer screen does not make the process of analysing the records any different.

The next logical step is to capture the data in a computing system that would enable searches to be performed on the records. It was therefore decided to construct an application that would not only hold the data and allow searches on it, but also one that would enable the records to be shared without them having to be sent as data on request. To this end a Web application would not only enable analysis of the data through a computing system, it would also provide global availability to authorised users, solving the problems associated with sharing paper and/or scanned documents.

Treating the data as meta-data raises its own challenges. When analysing the data set, it became clear that some parts of the form were rarely and/or partially used and as such may require adjustment or modification rather than simply converting the paper version directly into database fields. One such example of this is a section on pathology that was used in just two records, and then only partially, meaning that it could be summarised into a single heading for the database, rather than a series of separate section that would go mostly unused.

Although an electronic recording system is available for osteological data (Osteoware, published by The Smithsonian Institute), this was judged to be unsuitable for use with mummified remains. Mummies rarely provide precise bone osteometric data, for example, unless the preserved soft tissue is stripped from the body or measurements are obtained from CT data. In addition, Osteoware does not include provision for recording data related to soft tissue preservation or to associated grave goods and other artefacts which were recorded for the Dakhleh Oasis mummies.

The recording of dental records presents another important issue, as different records used different numbering systems. Although this was easily accounted for by simply placing both numbering schemes on the input labels, the more important underlying questions are how much data must be recorded, and is it prudent to change recorded data from one system to another for the sake of consistency? The latter is especially important as although it would make data mining simpler, it introduces an additional risk of human error.

Selecting an application architecture

There are three basic computing system architectures employed for holding records, i.e. databasing, a standalone application with its own data installed on a single user's machine, a client application that can be installed on any machine that connects to a single shared data source, or a completely on-line application with its own data that any user can access.

An on-line application has several advantages over a standalone application. Updates to applications, particularly in fields such as archaeology where the adoption of IT recoding methods is nascent, are necessary and can be frequent. Although an application can connect to the internet to check if an update is necessary, it would not guarantee that all users of the system are using the same version. The biggest disadvantage is that the data is not shared other than by means of replication if the application is isolated with its own data. A totally on-line application has the disadvantage of not being available if it is not reachable, due to connectivity issues, local or otherwise, or if the application's server is not functioning as required.

Having a standalone application, provided to all who wish to use the system, linking to shared data represents the worst of both scenarios, i.e. users need to be on-line (operating on locally cached copies of data poses its own challenges with regards to synchronisation) and that any fundamental changes e.g. to the data structure, could result in out of date versions of the software not being able to connect to the data source, or worse still, cause erroneous data to be received and/or recorded.

Internet connectivity is now taken for granted, as it is now even available on the common mobile phone. Therefore the expectation that machines used by archaeologists are connected to the Internet, at least in the office if not in the field, is a reasonable assumption to base the choice of which architecture to employ. The use of an Internet based only application, despite possible issues of availability under certain circumstances, is hence the better of the three options.

Determining Requirements

Ideally, a web application designed for recording archaeological data would provide a simple user interface that does not require an understanding of the database's underlying mechanics, and would have a simple enough structure for non-experts to provide feedback on. However, traditional applications are by their very nature closed off to the user, and any facilities that are made available to users require programming knowledge, which cannot be expected from those without a background in computing.

In its simplest form, the need is for a web application that will run within a browser on a computer or a tablet device that will transparently connect to a backend database. It is assumed that although any such application could be accessed on a small mobile device with a browser, i.e. a handheld smartphone, a version adapted for limited screen space was not considered at this stage due the limitations of using such a device to access detailed archaeological data.

The application's interface needs to be not only intuitive to use, it also needs to be easily adaptable to changes requested by users. Every application is technically adaptable, but the time (and therefore expense) required does not always make this feasible. Modifications will naturally require the services of a developer, but the ease with which a developer can effect such changes is an important consideration in a field where systems once developed are not only closed and isolated, but rarely updated.

Development platform selection

Web applications can be developed using any one of a multitude of available development tools. A combination of a development language and database would be required for a mummy database. The choice of an open database was made based on the most popular (Oracle Corporation, 2012) and hence best supported database, with MySQL best representing this criterion.

The selection of language and framework was a more complex task, as there are numerous environments available. Traditionally MySQL has been paired up with PHP, however in recent years new development platforms such as Ruby on Rails and Google Web Toolkit (GWT) have emerged that have been designed specifically for web application development (Clavijo, 2013).

These new platforms closely fulfil the requirement of facilitating the development of web based intuitive interfaces, and a data model based approach that allows for easier modification of an application than has been possible with traditional object orientated applications. The MVC (Model–View–Controller) (Leff and Rayfield, 2001) approach to applications development is one such platform model, where the data Model, user View, and user input Controller are separated, allowing for an application to be developed where each aspect can be tailored for different groups of users who would be working with the same set of data.

Ruby on Rails, or just 'Rails' for short, is a framework that is built on the Ruby language that is MVC based. Code re-use is also strongly encouraged and supported through the use of GitHub (GitHub Inc., 2012a) where developers are encouraged to share the code they have developed, resulting in a great number of program modules available for use when developing Rails applications.

As the application would be a record administration application also known as a CRUD (Create, Read, Update and Delete) application for working with ancient Egyptian mummy records, the best supported open source admin interfaces (i.e. CRUD's) available were evaluated.

There are over ten significant CRUD interfaces available as listed on Ruby Toolbox (Olszowka, 2012). Of these the three most popular were chosen for further evaluation, namely ActiveAdmin (179k downloads), Rails Admin (29k downloads), and Active Scaffold (no download statistics, but it ranked third in popularity). The evaluation for each of these interfaces was performed by building a simple test application for storing mummy records using the interface, and testing the resultant interface for basic flaws and robustness, i.e. validation checks etc.

Active Scaffold

This is an expansion for Rails' built in scaffold feature, for rapidly generating a basic application linked to a database, by issuing a single command line on a pre-configured environment (Anon., 2012b). Certain flaws in the framework were identified, and were unresolved at the time of evaluation e.g. attempting to create a record without a sub-record resulted in an inconsistent record being created. It does offer a visually simplistic interface, but this did not always work as expected when displaying data.

Rails Admin

This offered a more comprehensive admin interface that functioned well in that it was able to display different types of sub records correctly, i.e. a main mummy record, and an external examinations sub record. The interface itself seemed a little complicated in the way it dealt with records and sub-records, offering CRUD operations for both types on the same screen (GitHub Inc., 2012b).

ActiveAdmin

This is by far the most actively used interface at the time of evaluation and offered a substantial amount of documentation and simple guides provided by supporting developers. Its method for creating records and sub-records is clearly segmented so as not to cause an ambiguity when creating or deleting records, and/or sub-records. It also employs other Rails code modules, known as Ruby Gems, such as Formtastic (for input form generation) and Devise (GitHub Inc., 2012c) (for the full user management built into ActiveAdmin), that are supported, as with ActiveAdmin (Bell, 2011) itself via GitHub.

Development, testing and evaluation

The goal of the system was to record as much of the recorded data as possible on the form other than the freehand sketches. The system was built up record by record, with attention given to each field being added so that validations for the fields were specified in the data model, such as requirements for mandatory values and non-negative values for fields holding age, metric data etc. As the system is built in a framework, it will also constrain the creation, update and deletion of records to maintain database integrity, i.e. as each record is required to have a unique identifier, this is specified in the model, and no creation of a duplicate record sharing a unique identifier is then possible; the user is presented with an error informing them of this if such an attempt is made.

A master record containing the main details of the Dakhleh Mummy records, i.e. the first page, was created as a starting point. Following this the remaining attributes, such as dental information, external examination, internal examination etc. were added in due course as sub-records to complete the record set.

Testing the system, named MEPR (Mummy Electronic Patient Record) (figure 1) has been an interactive process that took place during development. After any major changes, some records would be created, updated, and deleted. The latter was done to test the validations for each record. Web connectivity and logins were tested by default on a regular basis as the system was updated on a live host and tested.

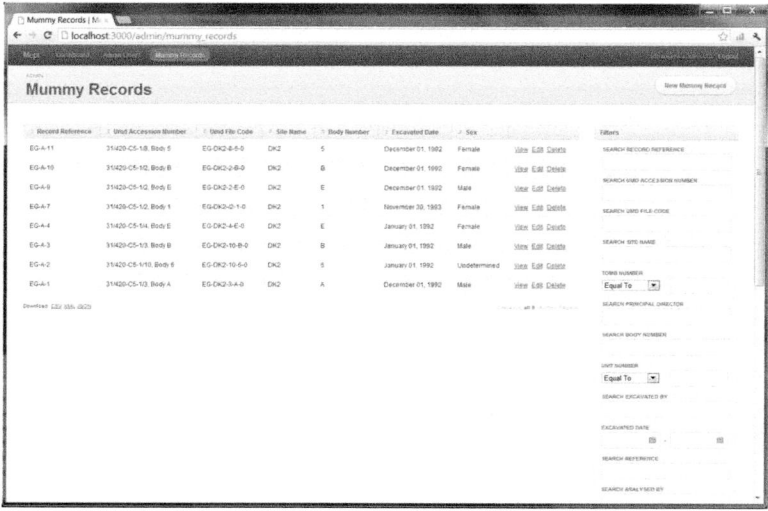

Figure 1: The MEPR main record list screen

One key feature is feedback driven improvements. Egyptologists are able specify which fields they wish to see in any record area, as it is effortless to implement such a change in a Rails application, it is just a matter of enabling or disabling a line of code, as appropriate, to affect the changes required. The iterative process of feedback and modification can continue as needed until the application is one that they are comfortable with, or if in the future, new requirements require a redesign of the input form. Obviously some changes are likely to be more complicated, e.g. dependency changes when a data item may need removal that other data items are reliant on.

Discussion

Advantages of using MEPR

Significantly improved record searches:
Searching the Dakhleh Mummy record set is now a greatly improved process compared to manually searching the paper copies. The search filters provided alongside the list of records allows for as many criteria to be specified as there are fields, e.g. all females whose estimated age is less than 20, with a date of excavation after the 20th of August 1992, excavated from the Kellis 1 site. To complete a similar search using the paper records would require the entire set to be examined, and with each record being 13 pages long (if dental data is present) with information on different pages, making this a very time consuming process. This is a far cry from entering the filters on a web form and then ordering a search to be performed, with results given within seconds.

Many systems require users to have some expertise in being able to query the database using some form of SQL (Structured Query Language). As this requires a basic knowledge of relational database theory, it is unreasonable to expect users who are not within the field of computing to make use of these systems directly. However, ActiveAdmin's search filters allows a complex, multivariate search to be built up quickly with a user-friendly interface, greatly increasing the power of the system for the vast majority of MEPR's target audience.

Increased longevity of data:
The MVC framework mentioned above adds to the longevity of the data by keeping it on-line. Once the data is stored within a contemporary relational database, such as MySQL in this instance, it can be transferred to other databases with ease. For example if Postgres was to become the leading open source database, then a switch to this would be an effortless transfer of data from MySQL. Even the Rails applications that access the data can be simply instructed, by means of changing a single line of configuration, to use a Postgres database instead, as Rails is largely database agnostic.

MEPR's first release was an on-line Heroku hosted installation, which actually uses a Postgres database as opposed to MySQL. Hence MEPR was at that time being developed using differing deployment and development databases, the ability to do this is a Rails feature.

As the dataset is largely static, keeping offline backups on an optical disc (such as a CD or DVD) would be advisable, as the risk of a hack attack destroying or corrupting on-line data is ever present. Although optical media has a limited guaranteed lifespan (Iraci, 2005), simply having a recorded process to migrate it onto new media regularly would easily ensure its safety.

Conclusion and future work

This project represents the use of modern web and computer data storage technologies to capture and record a very significant dataset that was previously difficult to gain access to. Its availability is now truly global. The architecture is extensible (James, 2010), as it has been developed in an applications framework built for that purpose. Future additions and modifications to the structure of data does not need code to be deprecated and/or significant new code to be written. Additions to the data model will accomplish most minor changes, with additions and/or modifications to the controller only being required if adding the new data to the model is not sufficient to capture and handle the new data.

The inclusion of non-Egyptian mummified remains is another area in which MEPR could be expanded. This would have to be managed carefully, so as to ensure that any addition of data does not replicate what is already present, as it is difficult if not impossible to 'undo' such changes (Stodder, 2012). This stems from the simple premise that if two parallel structures are in place for recoding a data set for example, dental records, it is possible that the data entered can straddle both structures. If an attempt was subsequently made to remove one of the structures, which could be required due to any number of reasons, e.g. one supersedes the other, then untangling such a set of 'straddling' records would pose a considerable challenge.

One area of research to be explored is the improvement of the reasoning applied to the data when it is queried, by means of making use of an ontology to work with the system. This however requires a much greater, deeper understanding of the meta-data that could only be obtained with significant input from archaeologists.

References

Anon., 2012a. *7th International conference of the Dakhleh Oasis Project*. [Abstract booklet pdf] Leiden University. Available at: <http://media.leidenuniv.nl/legacy/dop-2012-abstracts-per-day.pdf> [Accessed 8 February 2014].

Anon., 2012b. *ActiveScaffold*. [online] Available at: <http://activescaffold.com/> [Accessed 3rd August 2012].

Aufderheide, A.C. 2003. *The scientific study of mummies*. Cambridge: Cambridge University Press.

Aufderheide, A.C. and Rodriguez-Martin, C., 1998. *The Cambridge encyclopedia of human paleopathology*. Cambridge: Cambridge University Press.

Aufderheide, A.C., Cartmell, L.L., Zlonis, M. and Horne, P., 2003. Chemical dietary reconstruction of Greco-Roman mummies at Egypt's Dakhleh Oasis. *Journal of the Society for the Study of Egyptian Antiquities*, 30, pp.1-8.

Bell, G., 2011. *ActiveAdmin*. [online] Available at: <http://activeadmin.info/> [Accessed 3rd August 2012].

Clavijo, D., 2013. *A Web framework comparison: Matt Raible's opinion*. [online] Available at: <http://blog.websitesframeworks.com/2013/03/web-framework-comparison-matt-raible-opinion-138/> [Accessed 1st April 2014]

GitHub Inc., 2012a. *GitHub features and project management*. [online] Available at: < https://github.com/features/projects> [Accessed 3rd August 2012].

GitHub Inc., 2012b. *Rails Admin*. [online] Available at: <https://github.com/sferik/rails_admin/> [Accessed 3rd August 2012].

GitHub Inc., 2012c. *ActiveAdmin tools being used*. [online] Available at: <https://github.com/gregbell/active_admin#tools-being-used> [Accessed 3rd August 2012].

Iraci, J., 2005. The relative stabilities of optical disc formats. *Restaurator*, 26, pp.134-150.

James, J., 2010. *A fresh look at Rails and Ruby*. [online] Available at: <http://www.techrepublic.com/blog/software-engineer/a-fresh-look-at-rails-and-ruby/3108/> [Accessed 4th August 2012].

Leff, A. and Rayfield, J.T., 2001. Web-application development using the model/view/controller design pattern. In: *Enterprise distributed object computing conference, 2001. Proceedings, Fifth IEEE International*. IEEE. pp.118-127.

Mills, A.J., 2013. *Excavations in Dakhleh Oasis, Egypt*. [online] Available at: <http://artsonline.monash.edu.au/archaeology/excavations-in-dakhleh-oasis-egypt/> [Accessed 8 February 2014].

Olszowka, C., 2012. *Rails admin interfaces*. [online] Available at: <https://www.ruby-toolbox.com/categories/rails_admin_interfaces> [Accessed 3rd August 2012].

Oracle Corporation, 2012. *Market share*. [online] Available at: <http://www.mysql.com/why-mysql/marketshare/> [Accessed 2nd August 2012].

Stodder, A.L.W., 2012. Data and data analysis issues in paleopathology. In: A.L. Grauer, ed. 2012. *A companion to paleopathology*. Chichester: Blackwell Publishing Ltd. pp.339-356.

Mummy website and database

Barbara Zimmerman, Sukeerthi Shaga, Pavitra Kaveri Ramnath, and Sai Phaneendra Vadapalli

Department of Computing Sciences, Villanova University, Villanova, USA

Abstract

A lifetime of work with mummies by Dr. Michael Zimmerman resulted in 3000 microscopic slides and paraffin blocks being donated to The University of Manchester's KNH Centre for Biological Egyptology. This collection offers researchers a treasure trove of materials in which to study the prevalence and history of disease in ancient populations. The collection is housed in Manchester, UK with limited accessibility to the materials. To widen the accessibility and make the data available to researchers around the world, a database has been constructed with a website front end. The database and website, constructed at Villanova University, contains the description of each individual slide, photomicrographs where available and relevant publications.

Introduction

The first items that have been entered into the database are from Egyptian mummies. These include mummies from the site of Dra Abu el Naga, on the west bank of the Nile, across from the modern city of Luxor. The tomb of Nebwenenef, the first high priest of Ramesses II, was found to contain a large number of intrusive burials. Other entries are from a very early Christian period (ca. 200 CE) settlement, Kellis, in the Dakhleh Oasis of the Western Desert of Egypt. Data from several mummies in museums are also stored. The group of over 200 slides from the Dakhleh Oasis mummies gives us a glimpse of the usefulness of the full database. Researchers can request loans of slides of interest to their particular study. Future enhancements of the database and website will include additional materials from mummies North and South America, including Alaska.

Accessing the database

The database is accessed through the internet at: http://manchestermummy.comze.com/search.php

This brings the user to the main search screen seen in figure 1. There are 4 possible selection choices: Country, Site, Organ, and Diagnosis. From this screen, the user can search the entire collection by allowing all the defaults of an 'all' setting to stand. If the user wishes to, this search can be narrowed to a particular country such as 'Egypt' by pulling down on the Country selection. By leaving the rest of the selection fields as 'all', the entire collection of Egyptian specimens will be shown.

Further narrowing can occur by selection of a site. An organ, for example, 'Heart' can be selected. If 'Heart' is selected but other fields are left as 'all', each specimen involving the heart, no matter what country, site, or diagnosis, will be displayed. The same procedure can be used for a diagnosis. For example, all cases of cirrhosis will be displayed, by just selecting the diagnosis of cirrhosis and leaving all other fields as 'all.' The choices are restricted to the diagnoses listed.

Once the selection is made, the 'Go' button at the bottom is clicked. A list of slides will appear. The selection of Country as 'Egypt', Site as 'Dakhleh Oasis', and Organ as 'Liver' yields the search result seen in figure 2. Three slides met the search criteria. One row was chosen and the 'More' button clicked. Figure 3 shows some of the fields that are part of the selected slide; these include history, provenance of the site and any publications in which the slide is referenced. One or more photo micrographs are shown if taken.

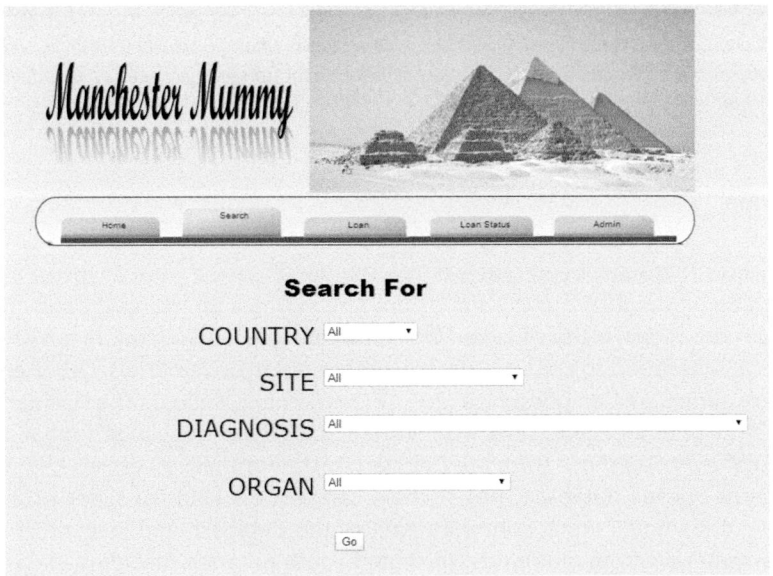

Figure 1: The main screen used for searching the database.

Loaning slides

Microscopic slides are available for borrowing. A request can be placed using the database by utilising a tab on the Search screen labelled 'Loan', as shown in figure 1.

The same screens are used until the very last screen. At the bottom the user clicks the 'Select Slide' button. This brings the user to a page that asks for 'name, institution, etc.'. The user fills out the information and submits the request by clicking the 'Submit Request' button at the bottom of the screen. The directors at the KNH Centre at The University of Manchester will communicate directly with the requester.

Manchester Mummy

Home | Search | Loan | Loan Status | Admin

You have selected the following:

COUNTRY: Egypt
nSITE: Dakhleh Oasis
DIAGNOSIS: All Diagnosis
ORGAN: Liver

Number of rows retrived is: 4

More	slide Id	AccessNum	Box Num	Sample Type	Organ	Diagonosis	Source Name
◉	99000095	D47-1	6	Microscopic Slide	Liver	Cirrhosis	Dakhleh Oasis
○	99000532	DO-36	26	Microscopic slide	Liver	No diagnosis	Dakhleh Oasis
○	99000543	DO-47	26	Microscopic slide	Liver	Cirrhosis	Dakhleh Oasis
○	99000544	DO-48	26	Microscopic slide	Liver	Cirrhosis	Dakhleh Oasis

More

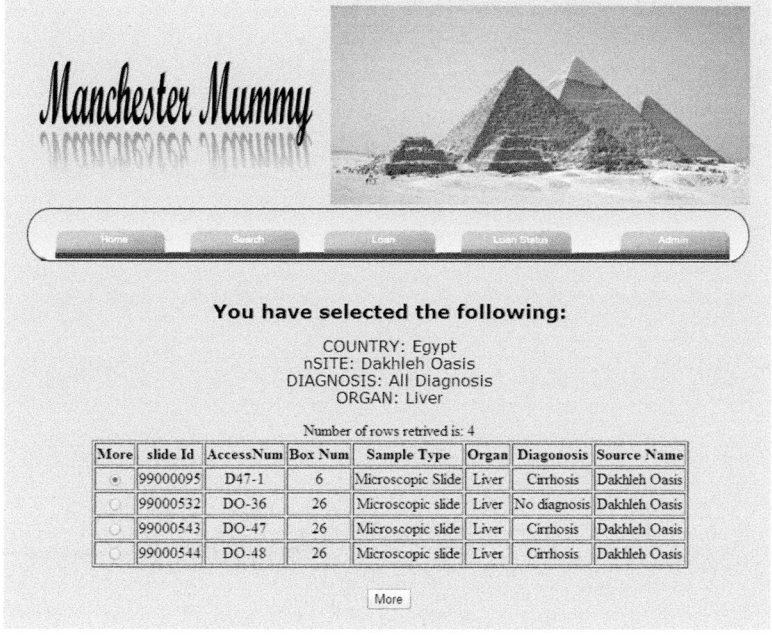

Figure 2: An example search result.

Detail of

99000543

slide Id	AccessNum	Box Num	Sample Type	Organ	Source Name	Notes	Size	Weight	Date Aquired	Natural/Art	Wrapper	Sex	Age	Date	Condition	comments	Specimen Name	Institution Name
99000543	DO-47	26	Microscopic slide	Liver	Dakhleh Oasis	1 slide, 2 Trichrome, 1 Victor			Mar-94	Artifical	unwrapped	Male	unk	200 AD	good	DO-38 through DO-57 is one individual	31/420-C5-1/2-M Tomb 2 Body M	Mount Sinai Hospital, NY, USA

Tests and Results

test	diagonosis	Details
Microscopic Examination	Cirrhosis	

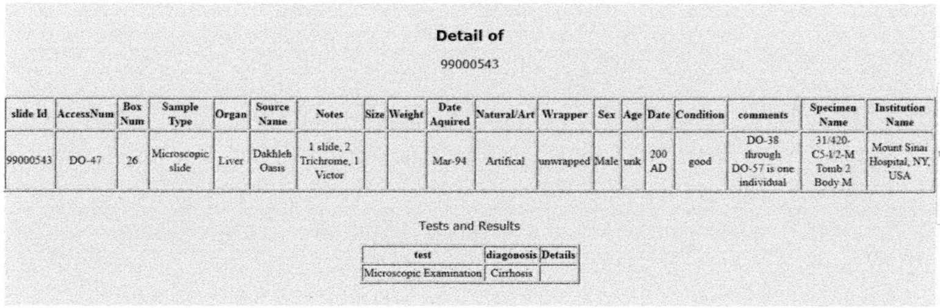

Figure 3: An example of the range of data available for a single slide.

Database structure

The database itself is implemented using open source software of PHP and MySQL. Open source means that the software can be downloaded from the Internet at no charge. The benefit of using open source software is that it is easily transferable to other institutions, making the code and structure used for the project useful on a very wide basis. The websites available to access downloads of the PHP and MySql software are http://php.net/downloads.php and http://dev.mysql.com/downloads/.

The authors hope that the database will prove useful to researchers and solicit comments for enhancements. Please contact barbara.zimmerman@villanova.edu or the KNH Centre for Biomedical Egyptology (http://www.knhcentre.manchester.ac.uk/).